Mines of *Battle Mountain, Reese River, Aurora and other Western Nevada Districts*

by J. M. Hill

This is a photographic reproduction of the 1915 U.S. Geological Survey Bulletin 594 entitled "Some Mining Districts in Northeastern California and Northwestern Nevada."

Coverage includes Lander, Mineral, Douglas, Lyon, and Washoe Counties in Nevada and Lassen and Modoc Counties in California.

Published by Stanley Paher

Nevada Publications
Box 15444
Las Vegas, Nev. 89114

ISBN 0-913814-64-4
Printed in the U.S.A.

CONTENTS.

ILLUSTRATIONS.

PREFACE.

By F. L. Ransome.

Scattered over its sparsely settled and generally arid expanse the State of Nevada contains approximately 200 centers of past or present mining activity. Some of these mining districts, such as the Comstock,[1] Eureka,[2] Tonopah,[3] and Goldfield,[4] have added millions to the world's wealth in precious metals and have been given thorough geologic study. Others of less note, including Silver Peak,[5] Manhattan,[6] Bullfrog,[7] Yerington,[8] Searchlight,[6] and Jarbidge,[9] have been made the subjects of brief reports by Federal geologists.

In 1908 Waldemar Lindgren, then in charge of the section of metalliferous deposits, recognizing the desirability of acquiring recent and authentic information about the numerous geologically undescribed districts in Nevada, instructed F. L. Ransome to visit a number of such districts in Humboldt County and W. H. Emmons to make a similar but more extensive examination of parts of Elko, Lander, and Eureka counties. The results of these reconnaissances were published in 1909 and 1910.[10] There still remained many widely scattered districts, some of them remote from regular lines

[1] Becker, G. F., Geology of the Comstock lode and the Washoe district, with atlas: U. S. Geol. Survey Mon. 3, 1882.

[2] Hague, Arnold, Geology of the Eureka district, Nev., with an atlas: U. S. Geol. Survey Mon. 20, 1892.

[3] Spurr, J. E., Geology of the Tonopah mining district, Nev.: U. S. Geol. Survey Prof. Paper 42, 1905.

[4] Ransome, F. L., The geology and ore deposits of Goldfield, Nev.: U. S. Geol. Survey Prof. Paper 66, 1909.

[5] Spurr, J. E., Ore deposits of the Silver Peak quadrangle, Nev.: U. S. Geol. Survey Prof. Paper 55, 1906.

[6] Ransome, F. L., Preliminary account of Goldfield, Bullfrog, and other mining districts in southern Nevada, with notes on the Manhattan district by G. H. Garrey and W. H. Emmons: U. S. Geol. Survey Bull. 303, 1907.

[7] Ransome, F. L., Emmons, W. H., and Garrey, G. H., Geology and ore deposits of the Bullfrog district, Nev.: U. S. Geol. Survey Bull. 407, 1910.

[8] Ransome, F. L., The Yerington copper district, Nev.: U. S. Geol. Survey Bull. 380, 1909.

[9] Schrader, F. C., A reconnaissance of the Jarbidge, Contact, and Elk Mountain mining districts, Elko County, Nev.: U. S. Geol. Survey Bull. 497, 1912.

[10] Emmons, W. H., A reconnaissance of some mining camps in Elko, Lander, and Eureka counties, Nev.: U. S. Geol. Survey Bull. 408, 1910. Ransome, F. L., Notes on some mining districts in Humboldt County, Nev.: U. S. Geol. Survey Bull. 414, 1909.

of travel, about which little geologic information was available, and in 1912 Mr. Lindgren directed Mr. Hill to make reconnaissance examinations of a number of districts in northwestern Nevada and of three districts in northeastern California. The results of this investigation are presented in this bulletin and in a paper on the Yellow Pine district, in Clark County, published in Bulletin 540.

During the year 1913 Mr. Hill made a similar examination of parts of eastern Nevada. When the results of that work are published and various reports now in preparation—namely, a monographic study of the Ely district, by A. C. Spencer; a report on the National district, by Waldemar Lindgren; a report on the districts of the Carson Sink quadrangle (Rawhide, Wonder, and Fairview); and a brief paper on the Rochester district, by F. C. Schrader—are issued, there will be few mining districts in Nevada for which at least some preliminary geologic information is not available.

Mr. Hill's reconnaissance had two main objects—(1) to satisfy the demand of the public for reliable information and (2) to gather data which should be of use in planning further geologic work in Nevada and in preparing a general report, for which full arrangements have not yet been made, on the geology and ore deposits of the State. Its purpose was thus directly practical, and there was little opportunity for Mr. Hill to engage in problems of broad geologic interest. Within the restrictions set by the nature of his task his observations have been carefully and accurately made and have brought out many facts of scientific interest, such as the widespread occurrence of adularia in the veins of northwestern Nevada, and the presence of secondary tetrahedrite and wurtzite in the Reese River district and of selenium in the veins of Aurora. His work thus promises to be of value not only to those interested practically in the ore deposits of Nevada but also to students of ore deposits who may be in search of trustworthy observations by which to test or amplify conclusions drawn from their own experience.

SOME MINING DISTRICTS IN NORTHEASTERN CALIFORNIA AND NORTHWESTERN NEVADA.

By J. M. Hill.

FIELD WORK AND ACKNOWLEDGMENTS.

From the first of July until the last of September, 1912, the writer made short visits to 16 camps in Mineral, Douglas, Washoe, and Lander counties, Nev., and to three camps in Lassen and Modoc counties, Cal. Owing to the large size of the area covered only a short stay could be made in each district, but the time spent was sufficient to afford at least a general understanding of the main geologic features of the district and of the nature of the ore deposits. The conclusions reached are not final, and in a number of the districts detailed work will be required to arrive at a satisfactory interpretation of the occurrence of the ores.

The map of Nevada north of latitude 39° 30′, as prepared by the Fortieth Parallel Survey, affords the best geologic data obtainable on the camps visited. This map shows the geology of the Peavine and Cottonwood districts in Washoe County and of the Battle Mountain, Skookum, and Reese River districts in Lander County. The so-called Archean rocks shown on this map at Peavine Mountain, north of Reno, are now known to be granites that are intrusive into pre-Tertiary schists. Battle Mountain consists in part of sediments older than the Weber quartzite, though it includes also the rocks of which it is shown to consist on Map V of the Fortieth Parallel Atlas. The determinations of many of the flow rocks given by Zirkel have been changed, largely owing to the development of the science of petrography.

The writer wishes to express his indebtedness to the mining men of the camps visited for numerous courtesies, especially to Mr. Pepper, of Mina; Mr. Cain, of Bodie; Mr. Deleray, of Pine Grove; Mr. Jensen, of Gardnerville; Mr. Stone, of Hayden Hill; Mr. Guyot, of High Grade; Mr. Liddell, of Battle Mountain; and Messrs. Marshall and Watt, of Austin.

In the preparation of this report the writer had invaluable advice and assistance from Mr. Waldemar Lindgren and Mr. F. L. Ransome. Many problems regarding the mineralogy and petrography of his collections have been solved by the cooperation of his associates in the Survey, to whom he wishes to express his appreciation.

ITINERARY.

From Mina and Luning, on the Goldfield branch of the Southern Pacific Railroad (see Pl. I), trips of five days and three days, respectively, were made into the Silver Star and Santa Fe districts. One day was spent at the Granite district northwest of Walker Lake, and a day at the Lucky Boy mine in the Hawthorne district at the south end of the lake. From Lucky Boy the stage was taken west to Aurora, in which camp three days were fully occupied. The 21st of July was spent driving down the valley of East Fork of Walker River to the old camp of Pine Grove, where one day was spent in the mines. Two and one-half days were spent visiting the mines on the west side of the Pine Nut Range, as far north as Red Canyon. From Gardnerville, in Carson Valley, the mines in the vicinity of Galena Peak of the Pine Nut Range were visited. From Reno a trip was made on the Nevada-California-Oregon Railway to Doyle, Cal. Two days were spent in an unsuccessful attempt to get into the Fox Mountains to see the old mines of the Cottonwood district, though the trip was not entirely in vain, for it was found that there is no metal mining in three of the districts shown on Plate IX of Bulletin 507,[1] namely, Round Hole, Deep Hole, and Sheephead.

On returning to Doyle the writer went to Hayden Hill, Cal. Two days were spent at the old camp of Hayden Hill, and the trail was followed northward through Adin to the Winters district, though only about a day could be given this small area. The trip from Winters northeastward up Pit River, past the hot springs to Alturas, took the greater part of a day. Five days were spent at High Grade, which at this time (August, 1912) was in the activity of a "boom." From High Grade the writer returned to Reno, spending three days in visiting some of the mines in the nearly deserted Peavine district. A week was spent in visiting the mines at Galena, Bannock, and Copper Basin, near Battle Mountain. Austin was reached on August 28, and four days were spent in a fruitless endeavor to get into the important caved mines and in examining some of the smaller mines on Lander Hill and in New York Canyon in the Reese River district. The next four days were spent in a trip to Kingston, an old camp on the west side of Smoky Valley at

[1] Hill, J. M., The mining districts of the western United States: U. S. Geol. Survey Bull. 507, 1912.

the extreme south end of Lander County. A trip back to Austin was made to visit the Skookum district, where about eight hours were spent, and the following four days were passed in visiting several mines on the east side of the Toyabe Range, the trips extending as far south as Washington and San Juan canyons, on the line between Lander and Nye counties.

GEOLOGY.

GENERAL FEATURES.

As will be seen from an inspection of Plate I, the camps visited by the writer are widely scattered, and as the greater part of his time was consumed in the study of the mines in small areas, he can contribute little to the knowledge of the regional geology. It is his belief that this paper will be of most value for its description of individual districts, and for this reason the following outline of the geology and statement of the nature and occurrence of the ore deposits as a whole are made as brief as possible.

In the region covered by this reconnaissance sedimentary rocks ranging in age from Paleozoic to Mesozoic were noted. These sedimentary rocks, in practically every camp in which they were noted, have been intruded by granitoid rocks, which differ somewhat from place to place, even in a single district, but which in spite of the local differences have a remarkably constant mineralogic composition throughout the area.

Flows of lavas cover a considerable part of the region visited, particularly along the western border of Nevada and in northeast California. Farther east, in Lander County, volcanic rocks are not so conspicuous.

In a few places in the western part of the area shown on Plate I, especially in northwest Mineral County and along Truckee River near Reno, there are thick accumulations of partly consolidated gravels, in some places interbedded with layers of volcanic ash, which are older than the deep Quaternary gravel fillings of the wide desert valleys that are so characteristic of the Basin province.

SEDIMENTARY ROCKS.

PALEOZOIC ROCKS.

The oldest rocks seen in this reconnaissance are the white quartzites of the Toyabe Range, which are exposed at the head of the San Juan Canyon. These quartzites may be of Cambrian age, though the correlation is uncertain. Several thousand feet of dark schistose shales that may possibly be of Silurian age overlie these rocks conformably and are also seen in the Skookum district. The lower

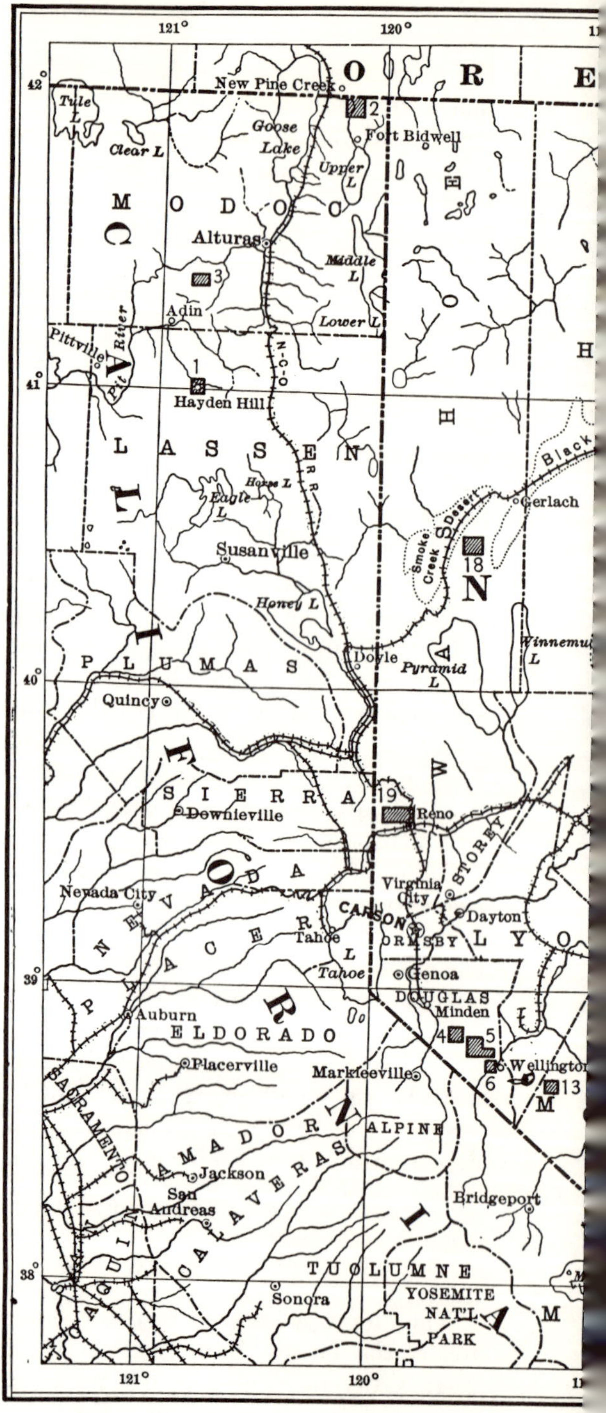

MAP OF A PORTION OF NEVADA AND CAL

DESCRIBED

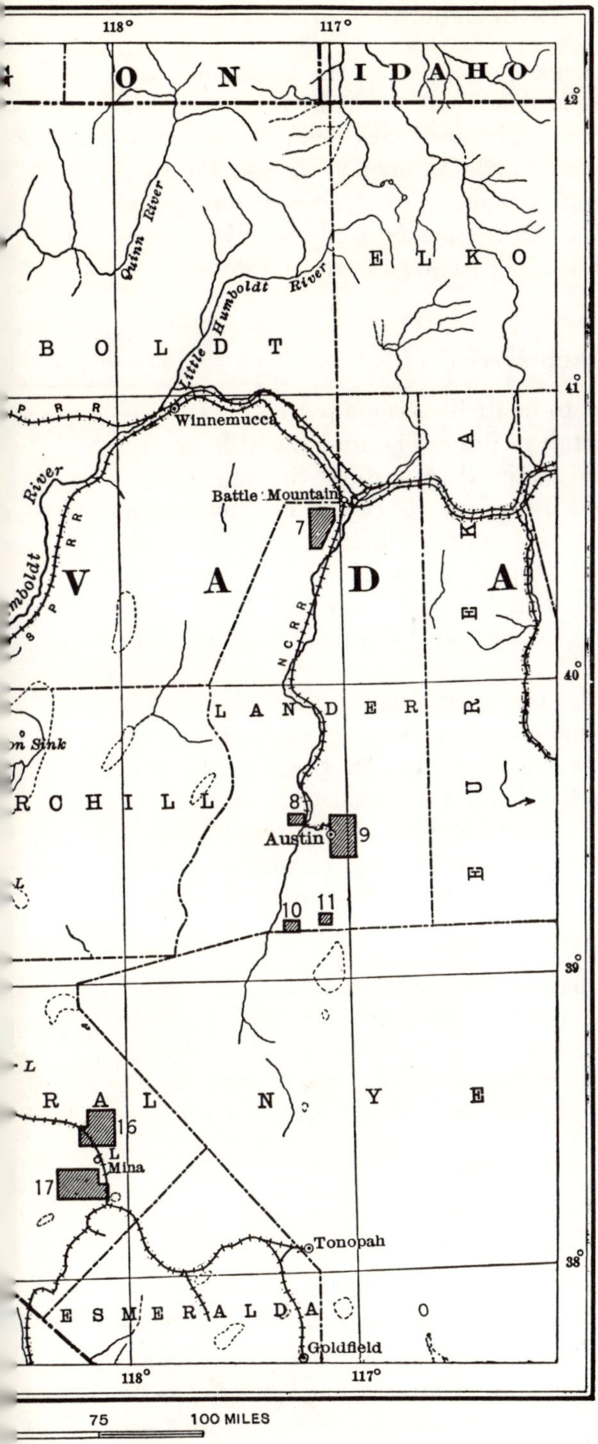

SHOWING LOCATION OF MINING DISTRICTS
REPORT.

Carboniferous (Mississippian) is represented by the limestones seen on both flanks of the Toyabe Range, south of Austin, and by the highly indurated dark slates and light and dark colored sandstones and quartzites of the Toyabe and Battle Mountain ranges; Pennsylvanian limestones overlie the sandstone and quartzites at two places in the Battle Mountain Range. Ore deposits occur in all of these formations with the exception of the Pennsylvanian limestones.

MESOZOIC ROCKS.

TRIASSIC SYSTEM.

Though not certainly known to be of Triassic age, the great accumulations of light-colored crystalline limestone and the still greater thickness of water-laid volcanic material seen along the entire western border of Nevada from the Silver Star district to Peavine Mountain are thought to be the equivalents of the Star Peak and Koipato formations characteristically developed in the Humboldt Range.

White crystalline limestones, which are thought to be of Star Peak (Triassic) age, were seen in the Santa Fe, Lucky Boy, and Red Canyon districts.

In the Silver Star district the rocks are largely bedded andesitic agglomerates, ranging from fine argillitic-looking rocks to coarse conglomerates of dark-gray to white color. In the Red Canyon district what is thought to be the Koipato formation is a series of andesitic conglomerates. On Peavine Mountain the rather schistose, dark-colored andesitic arkoses are thought to be of similar age. Ore deposits are found at various places in all of the supposed Triassic rocks.

JURASSIC SYSTEM.

Reddish-colored conglomerates, calcareous sandstones, and limestones of Jurassic age are exposed in the Pilot Mountains at the south end of the Silver Star district. So far as seen by the writer these beds are not mineralized, though some of the mines at the south end of the range are probably inclosed in Jurassic rocks.

CENOZOIC ROCKS.

TERTIARY SYSTEM.

PLIOCENE SERIES.

Partly consolidated gravels, silts, and sands form prominent cliffs along Truckee River west of Reno, and in the valley of East Walker River between Pine Grove and Aurora. In this valley several beds of white rhyolitic ash and lenses of what appeared to be diatomaceous earth were noted. In both places there are thin beds of poor coal. A small quantity of this coal has been mined for purely local use by ranchers along East Walker River at a place near the bridge on the Mason Valley-Hawthorne road.

QUATERNARY SYSTEM.

The broad flat intermountain valleys of the region are filled to an unknown depth with accumulations of largely unconsolidated gravels, sands, and silts. Evenly sloping alluvial cones extend from the mouths of the larger canyons for long distances toward the center of most of the valleys. The material composing these cones is coarse near the mountains but gradually diminishes in size of grain to the finest sands and silts, which are often wind borne and form dunes about the flat, absolutely barren playa lakes which occupy the central parts of many of the valleys.

IGNEOUS ROCKS.

INTRUSIVE ROCKS.

Granitoid rocks were found intrusive into the sedimentary rocks in the Gardnerville, Red Canyon, Battle Mountain, Reese River, Washington, Kingston, Hawthorne, Silver Star, and Peavine districts.

These rocks range from diorites through monzonites, quartz monzonites, and granodiorites to granites. The diorites generally carry some quartz and orthoclase, and the granites as a rule show some of the plagioclase feldspars, so, although the range appears to be large, in reality it is rather small. The quartz monzonite of Austin, Lander County, is very similar to the quartz monzonite mass near Wellington at the south end of the Pine Nut Range, in Douglas County, and to the intrusive rocks of Peavine and Fox mountains. These intrusives vary from granular to porphyritic in texture.

As a general rule the larger masses are granular, but many carry large pink crystals of orthoclase, whereas the smaller dikelike bodies are porphyritic, showing phenocrysts of both orthoclase and plagioclase feldspar, together with biotite and hornblende. Usually in the porphyritic phases the quartz is carried in the groundmass of the rock, though in the Battle Mountain district there are dikes of plagioclase-bearing porphyry which have large, well-developed quartz phenocrysts. At the mouth of Birch Creek, about 16 miles southwest of Austin, the normal quartz monzonite grades into a coarsely porphyritic alaskite porphyry, consisting almost entirely of quartz and feldspar, and a mass of similar rock is seen north of the road to the Granite district, at the northwest corner of Walker Lake. In most of the large bodies of this monzonitic rock there are smaller dikelike segregations of basic and acidic rocks. The aplitic dikes were noticed particularly in the Wellington district, though both aplitic and lamprophyric dikes were present. The latter type is more common at Austin, though aplitic dikes are present.

The intrusion of this type of rock is believed to have caused the mineralization in the districts in which it is found. Ore deposits occur in the "granite" in the Austin, Peavine, Red Canyon, Wellington, Granite, and Pine Grove districts.

The age of the intrusive was not determined at any place, though they cut Jurassic rocks; but it is thought that they were all injected during the late Cretaceous or early Tertiary period of intrusion common to the Sierra Nevada and Great Basin provinces.

EXTRUSIVE ROCKS.

As is well known, volcanic rocks are widely distributed throughout the Great Basin province. In each district visited by the writer volcanic rocks were present, though in several districts they are not of great areal extent. The ore deposits of the following districts are inclosed in volcanic rocks of probable Tertiary age; Hayden Hill, Winters, High Grade, Gardnerville, and Aurora. The ore deposits of the eastern end of the Peavine district occur in volcanic rock.

In northeastern California the older volcanic rocks are both rhyolitic and andesitic in composition and are capped by basaltic flows. In almost all the districts visited in Nevada augite andesites are the prevailing volcanic rocks of economic interest, though biotitic andesites and rhyolites and basalts are present at different places.

ORE DEPOSITS.

The ore deposits discussed in this report may be grouped under the following headings: Gold-silver deposits, silver-lead deposits, copper deposits, and antimony deposits. Each of these broad groups may be further subdivided according to the mineral composition of the ores and their occurrence.

AGE.

It is the belief of the writer that, with the exception of the few deposits of placer gold, the metallization of all the districts visited occurred within two or possibly three of the metallogenetic epochs distinguished by Lindgren.[1]

LATE TERTIARY DEPOSITS.

The latest mineralization of the region took place after the outpouring of the Tertiary andesitic and rhyolitic lavas, and was par-

[1] Lindgren, Waldemar, Metallogenetic epochs: Canadian Min. Inst. Jour., vol. 12, pp. 102–113, 1910.

ticularly productive of gold and silver bearing veins associated in some places with selenium and tellurium.

The following districts, described in this report, were mineralized during this epoch: Hayden Hill, Winters, and High Grade districts in California, and Gardnerville, Aurora, and parts of the Peavine and Silver Star districts in Nevada.

EARLY TERTIARY (?) DEPOSITS.

The older deposits occurred after the intrusion of the granitoid rocks, the age of which is not definitely known, but it is believed to be either late Cretaceous or early Tertiary. As Lindgren has shown, metallization took place during both epochs, so there is some question as to which of his divisions includes the majority of the ore deposits described in this report, yet it seems probable that most of the mineralization took place after the post-Jurassic intrusions and folding, for, as a rule, silver, lead, and copper are important in these deposits, which Lindgren states is not the case in the ore deposits formed in late Mesozoic time.

In the Red Canyon, Wellington, Battle Mountain, Skookum, Reese River, Washington, Kingston, Granite, Pine Grove, Hawthorne, Santa Fe, and parts of the Silver Star and Peavine districts the ore deposition probably took place in early Tertiary time.

GOLD-SILVER DEPOSITS.

DEPOSITS OF THE EARLY TERTIARY (?) EPOCH OF METALLIZATION.

Gold-silver deposits which seem to have been formed during the early Tertiary epoch of mineralization are found in the Peavine, Red Canyon, Wellington, Pine Grove, Granite, Santa Fe, Silver Star, and Battle Mountain districts.

In the Peavine, Red Canyon, Santa Fe, and Silver Star districts the gold deposits are veins in sediments of supposed Triassic age. These veins are usually simple fissures filled with crushed and altered country rock, and white quartz carrying pyrite, chalcopyrite, and more or less galena and sphalerite. The veins of the Wellington and Granite districts are similar in mineralogic composition but occur in quartz monzonite. The gold deposits of the Battle Mountains are likewise quartz pyrite veins, but they occur in sediments of Paleozoic age.

In all these gold-silver veins the ratio of the sulphides varies within wide limits. The metallizing solutions, however, in most places affected the wall rocks in a similar manner, producing zones of softened sericitized rock near the veins.

The Pine Grove deposit deserves particular mention, as it represents a deposit of different character from that of most of the deposits seen during this reconnaissance. The mineralization is quite similar, consisting of quartz, pyrite, and chalcopyrite, but the occurrence is different from that of most of the deposits. The ore occurs in flat north-dipping lenses in dark, highly biotitized quartz monzonite through a width of 150 feet. A somewhat similar though much less extensive alteration was noted along the Hidden Treasure vein in the Santa Fe district, which also occurs in altered quartz monzonite. In this vein, however, a small amount of galena is present.

In practically every camp where gold veins of the supposed early Tertiary period of mineralization occur there are also more or less extensive placer deposits.

DEPOSITS OF THE LATE TERTIARY EPOCH OF METALLIZATION.

Two somewhat distinct yet overlapping types of ore deposits were formed after the outpouring of the andesitic and rhyolitic lavas. In both types the country rock in the vicinity of the veins has suffered intense alteration, in some places being hardened by the addition of silica and in others being softened by a sericitization of its constituents.

The gold-bearing veins of High Grade, Winters, Hayden Hill, Aurora, and the veins in the vicinity of Silver Star in the Silver Star district belong to one class of the late Tertiary deposits. In all these districts the deposits consist of simple veins with numerous offshoots into the highly altered country rock. The gangue minerals are very fine grained white quartz with adularia, a secondary feldspar, and more or less calcite. In all of them the quartz has replaced calcite to some extent, giving a peculiar platy structure to the quartz, as shown in Plates VIII (p. 50) and XVII, *B* (p. 149). With the exception of the veins near Aurora, the veins themselves carry almost no sulphides, though pyrite is found in the altered wall rocks adjacent to the veins. The gold is all free and is usually so finely distributed that it can hardly be seen with the unaided eye. At Aurora besides the free gold the ores carry tetrahedrite and some pyrite and chalcopyrite. They also contain selenium, though the nature of the selenide compound is not as yet known. The gold bullion from these ores is very high in silver, being worth only $8 an ounce.

The pay ore in these quartz-adularia veins occurs in shoots and, so far as known, is not equally distributed throughout any one vein. The richer ores of the districts described in this report occurred within 350 feet of the surface, though developments do not show that they do not continue below that level, except in the Hayden Hill district, where the ores in the shoots have been of about equal grade throughout.

The deposits at the eastern end of the Peavine district, sometimes called the Wedekind, in the vicinity of Gardnerville, and a few of the deposits at the eastern end of the Silver Star district represent the second class of late Tertiary mineralization. At these places the veins show no secondary development of quartz after calcite. In fact the ores occur along zones of intense hydrothermal alteration, which, in the most productive mines, has caused the rock to become soft and light colored, owing to the almost complete sericitization of the feldspars and ferromagnesian minerals. Pyrite is abundantly disseminated in these bodies of altered rock and is seen in stringers, associated with a little galena and sphalerite, cutting the mass in all directions. The gold in these deposits appears to be free and at the surface there are usually pockets of rich material. With depth the grade of the ore seems to diminish, though as little work has been done on any of the deposits of this type the statement that all of them become worthless with depth is not justified.

As a rule the late Tertiary gold deposits do not produce placers, though the whole southwest side of Hayden Hill was worked for its placer gold in the early days of the camp.

SILVER-LEAD DEPOSITS.

So far as known, all the silver-lead deposits of the districts visited are closely associated with the intrusive quartz monzonite and associated rocks and are therefore of late Mesozoic or more probably early Tertiary age. Practically all the silver-lead ores are found in veins in sedimentary rocks of both Paleozoic and Triassic (?) age and more rarely in the intrusive quartz monzonite. In the Red Canyon, Silver Star, and Santa Fe districts the veins occur in sediments of probable Triassic age, whereas in the Battle Mountains, and in the Skookum, Washington, and Kingston districts, and at the north end of the Reese River district the veins cut sediments of Paleozoic age. The principal veins of the Reese River district occur in quartz monzonite and are the only well-developed silver-lead veins seen in the intrusive rocks in any of the districts visited.

In all these veins there is a marked similarity in the sulphide minerals, though the proportions of one to another differ widely from place to place even in the same vein. Antimonial compounds, generally the copper-silver tetrahedrite known as freibergite, are present in practically every vein. Galena, dark sphalerite, pyrite, arsenopyrite, and chalcopyrite are usually found to be more or less abundant.

Where these veins cut sedimentary beds there is usually not much apparent alteration of the rocks, though in most such places they have undergone slight sericitization. In the quartz monzonite at Austin the solutions that formed these veins have markedly bleached and sericitized the wall rock for short distances on either side of the vein.

At the Lucky Boy mine, in the Hawthorne district, the silver-lead vein cuts through contact-metamorphic limestones of Triassic (?) age and through the intrusive quartz monzonite, so it is thought, though with some question, that it was formed after the major alteration of the limestone.

In the silver-lead veins there are usually three well-marked and distinct zones or belts of ores. From the surface to the original ground-water level, which in the camps described in this report was usually within 100 feet or less of the surface, very rich "silver chloride" ores were found associated with lead carbonate. At and for a short distance below the water table there was in most mines a rather indistinct, usually narrow, belt of rich argentite ore. At different depths below the water level the zone of secondary sulphides was encountered. Polybasite seems to be the most common secondary sulphide in all the districts visited, and is associated with different quantities of stephanite and light and dark ruby silver.

In most of the districts visited the original sulphides lie relatively near the surface, though secondary mineralization occurs at different depths along fractures in the original ores. As a rule the original sulphides are medium-grade ores and the silver is associated with the tetrahedrite and galena. Stibnite is in places associated with the lead-silver ores in the Battle Mountains, though as a rule the deposits of workable antimony ore are distinct and do not carry much silver.

COPPER DEPOSITS.

DEPOSITS OF THE EARLY TERTIARY (?) EPOCH OF METALLIZATION.

The most important copper deposits are in the Peavine, Santa Fe, Hawthorne, and Red Canyon districts, where they occur in Triassic (?) sediments, and in the Battle Mountains and Washington districts, in sedimentary beds of Paleozoic age. In all these districts they are associated with quartz monzonite intrusives of probable early Tertiary age, and most of the deposits are of the contact-metamorphic type or are replacements in sandstones, andesitic arkose, calcareous shales, or limestones near masses of intrusive rocks. So far as developed the ores are in large part oxidized, though the original chalcopyrite characteristic of the copper deposits of this type is generally seen as remnants in the ores. Garnet and epidote are the most widely distributed contact minerals, but in the deposits of Copper Basin and Copper Canyon in the Battle Mountains the contact-metamorphic amphiboles and pyroxenes are associated with those minerals

DEPOSITS OF THE LATE TERTIARY EPOCH OF METALLIZATION.

The copper deposit of the Ruby Hill mine in the Gardnerville district is believed to occur in late Tertiary andesites, though there is considerable question whether the andesites may not be a part of the Triassic.

ANTIMONY DEPOSITS.

Deposits of stibnite ore and its oxidized products occur in the Battle Mountains at the Antimony King mine in Cottonwood Canyon and at two places near Big Creek, on the west flank of the Toyabe Range, about 12 miles south of Austin. At all these places the deposits occur in crushed, contorted, dark, siliceous shales, though at the Cottonwood Canyon locality the vein does pass into the overlying, highly indurated sandstone. The ores occur in lenslike masses, nearly if not quite parallel to the folded bedding of the shales, though they have a lodelike form. The ores in all places consist of white quartz and stibnite, but in some places they carry minor quantities of tetrahedrite and galena, showing that they are probably related to the silver-lead mineralization.

SUMMARY OF OCCURRENCES OF THE ORE DEPOSITS.

In the following table the attempt has been made to show in a graphic manner the types of rocks and the occurrence and character of the deposits in the various districts:

Summary of ore deposits of mining districts described in this report.

No. on Pl. I (p. 18).	Name of district and location.	Country rocks in which ore deposits occur.	Age of country rocks.	Types of ore deposits.	Age of ore deposits.	Chief metals produced.	Characteristic minerals of the deposits.
	CALIFORNIA.						
1	Hayden Hill, Lassen County.	Yellowish fine-grained rhyolitic tuff and coarse rhyolitic conglomerate.	Tertiary (?)	Fine-grained porcelain-like white quartz-adularia veins, and open brecciated veins.	Late Tertiary	Gold	Free gold, pyrite in altered wall rock.
2	Winters, Modoc County.	Dark-brownish and greenish basalt; some hydrothermal alteration.	do	Calcite and quartz adularia veins.	do	do	Do.
3	High Grade, Modoc County.	Dark brown and green andesite; yellow rhyolitic tuff; purple glassy flow rhyolite and basalt.	do	Brecciated country rock cemented by quartz and adularia. Pyrite in glassy rhyolite.	do	do	Free gold. Gold associated with pyrite.
	NEVADA.						
4	Gardnerville Douglas County.	Dark-colored augite andesites.	do	Quartz veinlets. Replacements in crushed altered andesite.	Late Tertiary(?)	do Copper	Free gold and gold associated with pyrite. Malachite, cuprite, chalcocite, and chalcopyrite.
5	Red Canyon, Douglas County.	Andesitic and rhyolitic conglomerates and arkose overlain by limestone, both intruded by quartz monzonite.	Triassic (?) Early Tertiary(?)	Veins in sediments. Contact-metamorphic deposits. Veins in quartz monzonite.	Early Tertiary(?)	Silver-lead Copper Gold-silver	Galena, dark sphalerite, and tetrahedrite. Copper carbonates, chalcopyrite, and epidote. White quartz and pyrite.
6	Wellington, Douglas County.	Quartz monzonite	do	Veins	do	do	White quartz, pyrite, some chalcopyrite, and galena.
7	Battle Mountain, Lander County.	Dark siliceous shale; light-colored sandstone, shale, and conglomerate; fine-grained white quartzite all cut by granite porphyry and quartz diorite; capped by andesite.	Paleozoic Early Tertiary(?) Tertiary	Veins in sediments. Contact-metamorphic deposits and associated replacements. Veins in andesite, local placers.	do Late Tertiary	Silver-lead, antimony, and gold-silver. Copper Gold	Galena, tetrahedrite, dark sphalerite, and pyrite; white quartz and pyrite, some chalcopyrite and galena, and stibnite. Copper carbonates, cuprite, chalcocite, chalcopyrite, sphalerite, and garnet. Free gold and pyrite.
8	Skookum, Lander County.	Dark siliceous shales	Paleozoic	Veins	Early Tertiary(?)	Silver-lead	Galena and tetrahedrite, sphalerite, and white quartz.
9	Reese River, Lander County.	Dark siliceous shale and quartzites. Quartz monzonite	do Early Tertiary(?)	Veins in shales and quartzite. Veins in quartz monzonite.	do	do	White quartz and rhodochrosite; galena, tetrahedrite, and sphalerite, chalcopyrite, arsenopyrite, and rich secondary silver sulphantimonides.

10	Washington, Lander County.	Dark siliceous slate, quartzite, and limestone cut by fine-grained quartz monzonite dikes (?).	Paleozoic	Veins in shales	do	Silver-lead	White quartz, siderite, galena and sphalerite, tetrahedrite and pyrite, and chalcopyrite.
						Antimony	Stibnite and white quartz.
			Early Tertiary(?)	Replacements in silicified limestone.	do	Gold	White quartz and pyrite.
						Copper	Copper carbonates and chalcocite.
11	Kingston, Lander County.	Dark siliceous slate, schist, limestone cut by quartz monzonite, quartz diorite, and rhyolite.	Paleozoic	Veins in sediments	do	Silver-lead	White quartz, galena, tetrahedrite, sphalerite, and rich secondary minerals.
			Early Tertiary(?)	Veins in quartz monzonite.		Gold-silver	Pyrite, chalcopyrite, and galena.
12	Granite, Mineral County.	Granodiorite	do	Veins in granodiorite	do	Copper	Copper carbonate, chalcocite, and pyrite.
		Andesite	Tertiary (?)	Veins cutting into andesite	Late Tertiary(?).	Gold-silver	Quartz, pyrite, and free gold.
13	Pine Grove, Mineral County.	Quartz monzonite, biotitized in ore zone.	Early Tertiary(?)	Replacement lodes	Early Tertiary(?)	do	White quartz, pyrite, and chalcopyrite.
14	Aurora, Mineral County.	Augite andesite	Tertiary (?)	Fine-grained porcelain-like white quartz veins.	Late Tertiary(?).	do	Quartz, adularia, calcite, free gold, tetrahedrite, and a little pyrite and chalcopyrite. Ore carries selenium.
15	Hawthorne, Mineral County.	Limestone and slate	Triassic (?)	Veins in sediments and intrusive.	Early Tertiary(?)	Silver-lead	Galena, tetrahedrite, sphalerite, pyrite, and chalcopyrite.
						Gold-silver	Quartz, pyrite, and chalcopyrite.
		Quartz monzonite	Early Tertiary(?)	Contact-metamorphic deposits.	do	Copper	Copper carbonates, chalcopyrite, and garnet.
16	Santa Fe, Mineral County.	Limestone	Triassic (?)	Contact-metamorphic and associated replacements in limestone.	do	do	Copper carbonates, chalcopyrite, some sphalerite, pyrite, and galena.
		Quartz monzonite	Early Tertiary(?)	Veins in sediments		Silver-lead	Galena, tetrahedrite, and pyrite.
				Veins in quartz monzonite.		Gold-silver	Pyrite, free gold, and some galena.
17	Silver Star, Mineral County.	Andesitic and rhyolitic arkose conglomerates cut by andesite dikes.	Triassic (?)	Fine-grained white quartz veins in sediments.	Late Tertiary	do	Quartz-adularia and free gold.
			Tertiary (?)	Quartz veins in sediments.	Early Tertiary(?)	do	Quartz, pyrite and chalcopyrite, and some galena.
						Silver-lead	Quartz, galena, tetrahedrite, sphalerite, and pyrite.
18	Cottonwood, Washoe County.	Slate (?) and quartz monzonite (?).	Triassic (?)	Veins (?)	(?)	do	(?)
			Early Tertiary(?)			Gold	Free gold and quartz.
		Volcanic rocks (?)	Tertiary (?)				
19	Peavine, Washoe County.	Fine-grained andesitic arkose and quartz monzonite.	Triassic (?)	Veins in sediments and intrusives.	Early Tertiary(?)	Copper	Copper carbonates, chalcocite, chalcopyrite, and pyrite.
			Early Tertiary(?)			Gold-silver	Pyrite and chalcopyrite.
		Augite andesite	Tertiary	Veins in andesite	Late Tertiary	do	Quartz, free gold, pyrite, and some galena.

MINING DISTRICTS.

In the following pages the mining districts in each county are described in order from north to south:

HAYDEN HILL, LASSEN COUNTY, CAL.

LOCATION AND ACCESSIBILITY.

The Hayden Hill mining district (No. 1, Pl. I, p. 18) is in the north-central part of Lassen County, Cal., near the forty-first parallel of north latitude and about 5 miles east of the one hundred and twenty-first meridian. This region is in the extreme southwestern corner of the area shown on the Alturas topographic sheet of the United States Geological Survey. The town of Hayden Hill is near the head of the west fork of Willow Creek, 16 miles south of Adin, at the west base of the low, rounded knob from which it is named. The nearest point on the Nevada-California-Oregon narrow-gage railroad is Likely, about 30 miles east-northeast of Hayden Hill. The mail route which in former times passed through the town started from Alturas and came by way of Adin, a distance of about 65 miles.

HISTORY AND PRESENT CONDITIONS.

Gold was discovered in the fall of 1870 by a party of clergymen en route from Susanville to Adin, who camped at the west base of the hill, near a small spring which issued from the Providence vein. Three of the four in the party were Charles Ellis, Barnes, and Haskins; the other's name could not be learned. These men had made camp for the night at this lonely water hole. Gold in the bottom of the spring attracted their attention, and naturally they began panning. It is said that they recovered from $250 to $500 a day by their united efforts.[1] The Providence vein was shortly discovered, the Hayden Gulch was second, and the Blue Bell ledge third.

In 1873 Raymond[2] reported that the Providence vein was an east-west fracture in sediments broken by volcanic action. This vein was from 1 to 6 feet wide, and had been developed to a depth of 75 feet. The ore was at first treated in arrastres and sluiced, but an 8-stamp mill was running by autumn. The gold, Raymond says, carried a lot of silver, was extremely fine, and much of it would float on water. The production in 1873 was estimated at $40,000.

---

[1] Raymond, R. W., Mining statistics of the States and Territories west of the Rocky Mountains, 1871, p. 55.

[2] Raymond, R. W., idem, 1873, p. 95.

During the following year, according to Raymond,[1] mining at Hayden Hill was at a standstill on account of the smallness of the veins. They were unquestionably rich, but the expenses in this remote district were too high to make them pay. It was found that all the soil on the north, west, and south slopes of the hill panned well.

By 1880 most of the veins now known had been worked to some extent, and the ensuing four years witnessed the greatest activity in the camp. During these years most of the soil on the southwest and south slopes was sluiced, the water being pumped from the east fork of Willow Creek.

Burchard[2] says of the period 1880–1884 that in 1880 the Juniper and Golden Eagle were the only mines in operation, but that the Brush Hill Mining Co. was installing a new water wheel. In 1881 the Providence and Golden Eagle mills were in operation, but in 1882 Juniper was the only producer.

The sedimentary rocks are completely brecciated. The gangue is a siliceous gravel, but the quartz does not carry the gold. The ores are screened and only the fine materials are sent to the mills. In 1882 there were five stamp mills and three arrastres working on East Fork of Willow Creek, 3 miles from the mines.

Preston[3] reports that in 1890 the faulted part of the vein at the west end of the Golden Eagle property had been found at the 160-foot level and that it yielded $30 ore. A rich chimney of ore was uncovered in the lower workings of the Brush Hill mine in the autumn of 1889, but the winter was so severe that all work had to be stopped until spring, when it required two and one-half months to unwater the lowest level.

The Golden Eagle, with several adjoining claims, was purchased by the Lassen Mining Co. in 1898, and for a few years following this purchase there was a revival of activity in the camp. From 1901 to 1910 the mine was worked for eight months each year, giving employment to about 120 men in the mine and mill. In 1910 a large part of the town of Hayden Hill was destroyed by fire, since which catastrophe very little work has been done in the district.

In the summer of 1912 only one of the old well-developed mines, the Golden Eagle, was open, and it is said that only by constantly replacing the timbers have the drifts been maintained. At the Brush Hill mine preparations were under way for reopening the shaft.

---

[1] Raymond, R. W., Mining statistics of the States and Territories west of the Rocky Mountains, 1874, p. 100.

[2] Burchard, H. C., Report of the Director of the Mint for 1880, p. 34; idem for 1881, p. 42; idem for 1882, p. 55; idem for 1884, p. 107.

[3] Preston, E. B., California State Min. Bur. Tenth Ann. Rept., pp. 275–276, 1890.

PRODUCTION.

There are no accurate figures of production for the Hayden Hill district. Approximate estimates for certain mines, furnished by residents of Hayden Hill, are given below.

Estimated production of gold and silver in the Hayden Hill district, Lassen County, Cal.

Juniper	$500,000– $600,000
Brush Hill	300,000– 400,000
Evening Star	150,000– 200,000
Hayden Gouge	20,000
Blue Bell	100,000
Providence	78,000
Hayseed	150,000
North Star	20,000
Golden Eagle (previous to 1901)	25,000
Golden Eagle (under Lassen Mining Co.)	1,000,000
	2,343,000–2,593,000

The following table was compiled by J. P. Dunlop, of the United States Geological Survey, from various sources. The figures from 1880 to 1901 are taken from the reports of the Director of the Mint, and those for some years include the whole of Lassen County. There were, however, during that time few producing mines in Lassen County outside of Hayden Hill. The figures since 1902, taken from the reports of the United States Geological Survey, have been collected with somewhat more care than were the mint figures.

Production of gold and silver from the Hayden Hill district, Lassen County, Cal.

Year.	Gold.	Silver.	Year.	Gold.	Silver.
1880 to 1886	$307,712	$26,285	1900	$19,807	$676
1887	39,109	303	1901	5,900	200
1888	50,000	200	1902	19,810	247
1889	....	....	1903	93,599	1,203
1890	14,890	200	1904	115,993	1,514
1891	3,676	....	1905	142,000	2,020
1892	15,400	....	1906	397,105	1,169
1893	....	....	1907	....	....
1894	35,283	....	1908	6,724	121
1895	25,000	....	1909	115,475	1,450
1896	40,300	....	1910	81,060	958
1897	49,100	850	1911	2,500	277
1898	37,460	300			
1899	28,898	....		1,646,801	37,973

As will be seen, there is a discrepancy of about $600,000 between the two sets of figures. This is due probably to the method of collecting statistics in the early days, and the larger figures are considered more nearly correct.

TOWN OF HAYDEN HILL, LASSEN COUNTY, CAL.

Looking N. 45° W. from Golden Eagle dump. Sky line shows even surface of the basalt-covered plains.

CLIMATE AND VEGETATION.

The summers in this part of California are pleasant; the days are moderately warm and the nights comfortably cool. Showers are common in summer. The winters are long and severe, and heavy falls of snow practically isolate camps like Hayden Hill from the first of December until the first of March. Late in winter the roads are barely passable, and not until the first of May can heavy freight be hauled into the camp. The precipitation is sufficient to fill numerous artificial lakes or tanks provided for the irrigation of the valleys.

The country between Susanville and Pit River is covered by a luxuriant growth of large long-leaf pines, and everywhere the plateaus are covered with fine forage. In most of the valleys there are hay ranches, as stock raising is the principal industry of this part of the country. The region gives great promise for the successful raising of the hardy fruits that have been planted of late years over large areas in Oregon, Washington, and Idaho.

TOPOGRAPHY.

The town of Hayden Hill lies at an elevation of 5,400 feet, at the west base of the hill of that name, which rises to a height of 6,357 feet. Seen from all directions this hill is rather conspicuous, rising about 900 feet above the relatively level basalt-covered plains. (See Pl. II.) On the south and west it has gentle even slopes to a point within 75 feet of the summit, which takes the form of a sharp peak. On the north and west it is very different, as branches of Willow Creek have cut in close to the hill. On the north, particularly, there is an abrupt rise of 150 to 300 feet near the summit, below which the slopes are very steep for another 300 feet. On the east side the hill rises 1,200 feet in a little over a mile with a rather flat bench 1,000 feet wide 500 feet below the summit.

GEOLOGY.

Hayden Hill is composed of buff-colored to yellowish rhyolite tuffs which look like sandstones. The bedding is distinct, being in some places horizontal and in others dipping very low to the west-southwest. These beds differ in appearance, some of them being composed of rather coarse conglomeratic material and others being fine-grained consolidated muds. In the conglomeratic beds the range in size of material is great, and the separate beds may be as thin as one-fourth inch. Some specimens of the rock show beautifully colored blendings of soft golden-brown, red, and purple tints, owing to the presence of iron oxides and manganese stains. At the summit of the hill these tuffs have been brecciated and silicified and cemented by white quartz with some adularia. This hard rock forms the knob or peak at the top of the hill and the scarp along the north side. The silicification and brecciation appear to have occurred in a fault that was slightly concave to the south-southeast and which has raised the beds to the southeast forming Hayden Hill. This fault zone strikes about N. 50° E., and the amount of vertical displacement seems to have been greater to the east-northeast near the summit of the hill than in the vicinity of the town of Hayden Hill.

Under the microscope the soft sandstone-like tuffs are seen to be composed of fragmental glass, containing here and there a few fragments of orthoclase. Quartz grains are very rare, but were found in one slide. The cementing material is a rather impure gritty kaolin. The rock from the summit of the hill is made up of fragments of silicified rhyolitic tuff in a matrix of white quartz and some adularia, most of which is granular, but crystals are developed in the larger spaces, having grown in from all the walls.

The low flat to the south and southwest of the hill is underlain in part by a soft, partly consolidated, dirty buff-colored rhyolite tuff, in which quartz and glass grains predominate, though some orthoclase and plagioclase feldspars are seen. The relations of this tuff to the formations of the hill are not clearly shown, though the unconsolidated tuff appears to rest unconformably on the slopes of the hill. A long tunnel on the southwest side of the hill penetrated this contact, to judge from the dump, but unfortunately the tunnel was caved about 800 feet from the mouth, and the part seen lay entirely in the soft, dirty buff rhyolitic tuff. About 2 miles north of Hayden Hill on the road to Adin, Willow Creek has cut deeply into soft greenish-white tuffs, which appear quite similar to those south of the hill. In this rock quartz, glass, and fragments that appear to be pyroxene are set in a kaolinized matrix.

The plains east and west of Hayden Hill were not visited, but those to the west are capped by a thin flow of dark rock that from a distance was thought to be similar to the basalt which covers much of the northeastern part of California.

It is said that at the bottom of the Golden Eagle shaft, 800 feet below the surface, there is a small exposure of porphyry. Only one small specimen of this rock was seen on the dump, and it was so weathered that it fell to pieces under the hammer. However, it seemed to be an andesite porphyry.

The exact age of the tuffs is not known. They unquestionably are of Tertiary age and were possibly formed near the later middle stage of volcanism, which lasted from the Oligocene to the close of the Tertiary. The basalts, if such they are, seen on the plains to the west of Hayden Hill are probably part of the flows of this rock which Mr. Diller has mapped in the northeastern part of the Lassen Peak quadrangle[1] and which he considers to be of Miocene or Pliocene age.

ORE DEPOSITS.

Location of veins.—The veins of Hayden Hill are confined to the area west of the summit, and the richest of them occur in the long even-sloping southwest side, as shown in Plate III. The most productive veins were the Providence, Golden Eagle, Juniper, Brush Hill, and Hayden Gulch. As shown on the map, there are two distinct sets of veins, one striking about N. 68° W. and the other N. 38° E. The northeast veins appear to be slightly younger than the northwest veins and certainly movement along the Providence has displaced the Golden Eagle, as at the northwest end of all the levels the Golden Eagle swings to the north on approaching the Providence. The two northeast veins near the main shaft of the Golden Eagle clearly cut that vein, but the post-mineral movement has been very slight.

Development.—The only mine that could be studied in detail was the Golden Eagle, and here the exposures were not good, as much of the vein had been stoped and could be seen only in a few pillars and at the ends of the stopes. The vertical shaft is said to be 800 feet deep, but the 300-foot level was the lowest which could be entered, as the shaft is in bad condition below that station. Water stands 625 feet below the collar. There are said to be no stopes below the 400-foot level. The drifts on the 100-foot level are 180 feet in length; on the 220-foot level, 1,100 feet; on the 300-foot level, 1,200 feet; and on the 400-foot level, 800 feet.

At the Juniper all the old workings are caved, but it is said that they were not over 200 feet deep.

The Brush Hill vein is developed by several shafts, two of which are equipped with whims. The workings were not accessible at the time of the writer's visit.

The Hayden Gulch vein is developed by a tunnel, which reaches a depth of about 75 feet but which could not be explored far on account of open underhand stopes. The older workings consisted of shallow shafts and a long series of open stopes that extend along the vein on the south side of the hill.

Structure of the veins.—As a rule the veins of Hayden Hill have steep dips. In general the northwesterly fissures dip 60°–70° NE., though at many places in the Golden Eagle mine the vein is vertical. The northeasterly fissures are as a rule nearly vertical, but at the junction of the Providence and Golden Eagle veins the Providence

---

[1] Diller, J. S., U. S. Geol. Survey Geol. Atlas, Lassen Peak folio (No. 15), 1895.

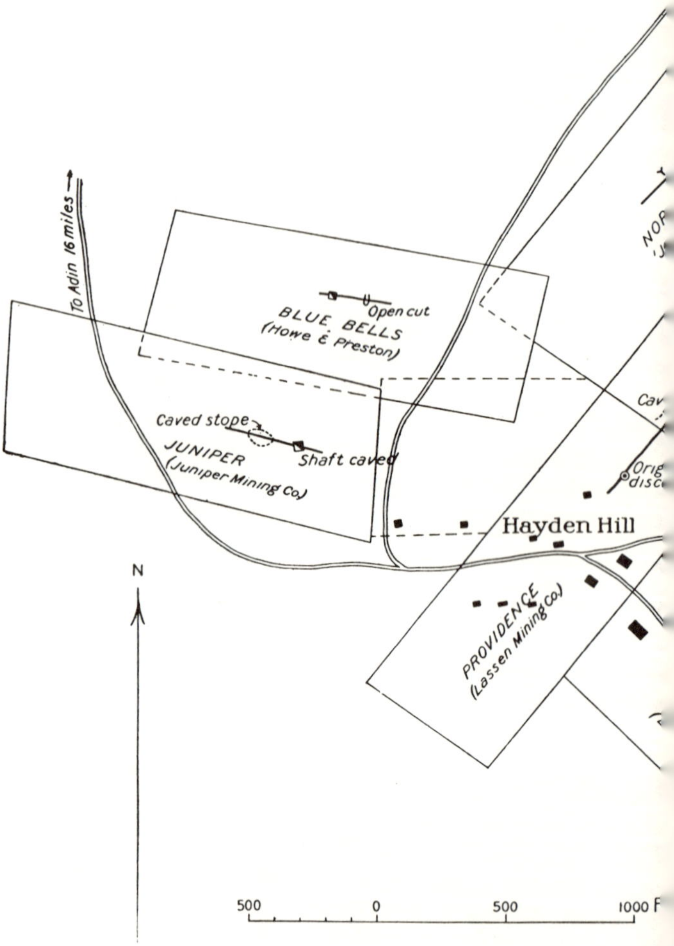

SKETCH MAP OF HAYDEN HILL, LASSEN COUNTY, C

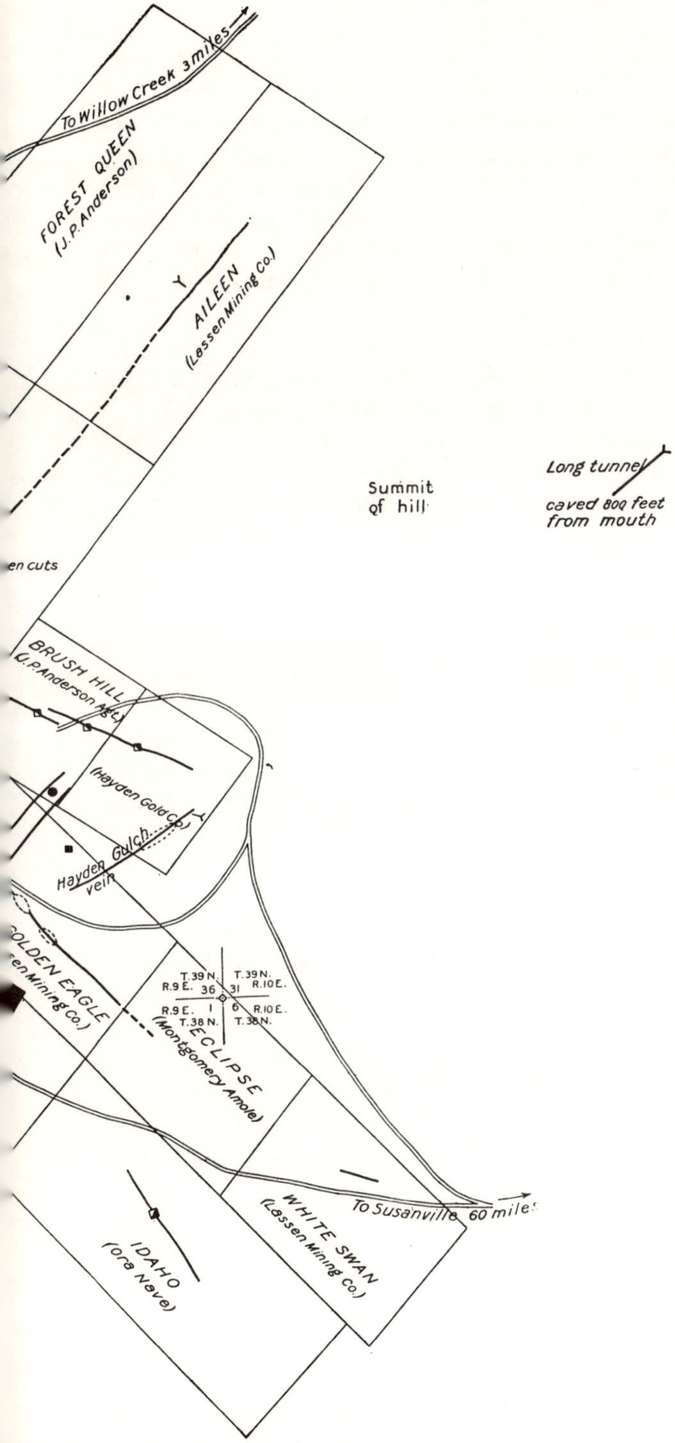

NG LOCATION OF THE PRINCIPAL CLAIMS AND VEINS.

has a dip of 45° SE., bringing the junction closer to the shaft with depth.

Most of the fractures are 1 foot to 10 feet wide, but some are as wide as 25 feet. They are not simple fractures, as is clearly shown by the branches of the Golden Eagle. On the 250-foot level of this mine there are two subparallel fissures southwest of the vein that dip less steeply to the southeast. One of these joins the main vein at the 300-foot level, but is 30 feet south of it at the 220-foot level.

Character of vein filling.—The filling in all the veins is a soft, unconsolidated fault breccia of the adjacent wall rock, much of the material being as small as sand grains. It will run freely, and it is said that almost no powder was used in working the veins. This tendency to run has made it necessary at the Golden Eagle to use much heavy timber to hold the drifts open. The same tendency has caused the caving of practically all the other mines. It is said that in one of the stopes in the Juniper mine the ore supply for a month was shoveled from one place, to which it ran of its own accord.

Both the country rock and ore are more or less iron stained, and the richest ore contains rather abundant pyrolusite. In fact at the Golden Eagle the "manganese" was always considered the highest-grade ore in the mine. This rich ore occurs in irregular pockets throughout the main vein, but appears to have been more abundant in the small fracture southwest of the main opening at the 220-foot level and at the junction just northwest of the shaft. Some ore from the Blue Bell shaft is a purplish-gray sand whose color is due to the manganese. The rich veins are remarkable for the small amount of quartz they contain. On the 220-foot level of the Golden Eagle shaft, 600 feet northwest of the shaft, there is a small lens of white sandy quartz that apparently contains no gold, and at other places some small bodies of similar quartz are seen.

Valuable constituents.—The gold occurs as small nuggets. It is not flattened, but is so fine that some particles will float when dried. It is never found in the fragments of rock, but always in the fine sandy, generally claylike material. This gold is really a combination of gold and silver worth about $14 an ounce.

Origin of the veins.—The only place on Hayden Hill where sulphides were found was in the Aileen tunnel, which runs southeast into the very hard siliceous scarp on the north side of the hill along the northwest fracture that dips northeast at very steep angles. The rock in this tunnel, which consists of white quartz and silicified rhyolite tuff fragments, is very dense, and the vein is small and hard. Both rock and vein contain a minor amount of adularia and pyrite. This ore is said to carry only small amounts of gold.

Origin of the shoots.—The concentration of gold with manganese in certain rather well defined shoots in the productive veins, and the fact that almost no sulphide minerals are present in them, lead one to conclude that the valuable ores are all secondary, owing to a concentration by oxidizing waters with a general downward movement.

The present ground-water level, as shown by the Golden Eagle shaft, is about 625 feet below the surface, and as all the workings studied were above this level, only the ores of the oxidized zone were seen, with the one exception noted.

It seems from the facts stated above that possibly the original veins of Hayden Hill were rather small quartz adularia veins carrying free gold and auriferous pyrite. The pyrite was possibly also disseminated to some extent in the porous wall rock near these veins. Later post-mineral movement crushed the veins and also a considerable portion of the adjacent walls, forming brecciated zones that ranged from 1 foot to 20 feet in width. The breccia thus formed an easy passage for the circulation of oxygenated surface waters which have leached the pyrite, leaving the gold scattered in the finer material of the breccia. The presence of manganese with the richer ore can not be satisfactorily accounted for, as no original manganese-bearing minerals were found. Its presence, however, is to be taken as a rather good indication that the ores in the deeper workings may be minable along certain channels, according to the hypothesis advanced by Emmons.[1] The veins are certainly younger than the rhyolite tuffs, which are of Tertiary age, and can be grouped with the late Tertiary deposits discussed by Lindgren.[2]

MILLING.

Owing to the fact that all the gold is free and is contained only in the fine material in the veins, crushing was early discovered to be unnecessary, so all the ore was dried and screened and only the fines were milled. The bullion produced was worth about $14 an ounce. The ore was originally milled in arrastres, 10 of which were in operation in 1884. Later there were several stamp mills 3 miles northeast of the mines, in the east fork of Willow Creek. In 1912 there was one stamp mill in Willow Creek using straight amalgamation methods and the Lassen Mining Co.'s mill, located near the Golden Eagle shaft. At the Lassen mill the ore from the mine was dried in a 10 by 20 foot Benjamin furnace, then put through a 1-inch

---

[1] Emmons, W. H., The agency of manganese in the superficial alteration and secondary enrichment of gold deposits in the United States: Am. Inst. Min. Eng. Trans., vol. 41, pp. 767–832, 1910.

[2] Lindgren, Waldemar, Metallogenetic epochs: Canadian Min. Inst. Jour., vol. 12, pp. 102–113, 1909.

mesh trommel, the oversize from which went onto the dump. It would seem as if the drying had not been complete, to judge from the amount of fines in the mill dump. The undersize from the trommel was again dried in two 3 by 10 foot revolving furnaces and crushed to 4-mesh size between rolls. It was again sized in a 4-mesh trommel, the oversize returning to the rolls and the undersize dropping to a storage bin. The sized pulp was leached in twelve 50-ton cyanide tanks. It is said that the rolls often produced "pancakes," or long slabs of partly dried ore, which stuck together because of the clay in them. These slabs, of course, were not easily leached. Another trouble was the large amount of slimes that was mixed with the coarser ore in the leaching tanks that prevented proper percolation of the solutions. On the whole, the mill as run was probably only about 50 per cent efficient.

HIGH GRADE DISTRICT, MODOC COUNTY, CAL.

LOCATION.

The High Grade district is located in the extreme northeast corner of Modoc County, Cal., in the Warner Mountains. (No. 2, Pl. I, p. 18.) Its northern boundary is the Oregon line; its eastern boundary is about 7 miles west of the Nevada-California line, and its western boundary is Goose Lake valley. Its southern limit is not yet defined, as there is a possibility of opening prospects south of the present known veins. Press notices in the mining journals indicate that during 1913 the veins near Willow Creek, 5 to 6 miles south of New Pine Creek, were being developed.

HISTORY OF MINING.

As far as can be learned, the discovery of gold in the High Grade camp was made on the Oregon claim at the south end of the district in 1905 by a sheep herder, who sold to J. O. Kafader, of Fort Bidwell. There is a story current that a scout by the name of Hoag, while stationed at Fort Bidwell about 1870, prospected in the Warner Mountains and brought in some gold. Before he could go back to his claims, however, he was killed in an Indian skirmish and all trace of his find was lost. Because of this supposed early discovery of gold in the district it was first called the Hoag district, which name was later changed to High Grade. The first local boom took place in the summer of 1905, but the autumn of 1909 and spring of 1910 witnessed the big rush to the district.

ACCESSIBILITY.

Until the spring of 1911 the nearest railroad was at Alturas, the seat of Modoc County, 52 miles south-southwest of the camp. In

1912 the Nevada-California-Oregon Railway had been extended along the east side of Goose Lake to Lake View. The town of New Pine Creek, 1 mile off this road, on the California-Oregon boundary line, is 8 miles west of the district and is the supply point for the miners. The roads from the town up Pine Creek to High Grade are in fair condition, though many steep grades and switchbacks are necessary to gain the top of the range. A triweekly mail stage operates between New Pine Creek and Fort Bidwell, going by way of the new camp.

High Grade is plentifully supplied with timber and water, though there is hardly enough water in the vicinity of the mines to generate power. It is an ideal summer camp, but it is said that from the middle of November until late in June the roads are practically impassable on account of heavy snows.

PREVIOUS DESCRIPTIONS.

The following is a list of papers published on the High Grade district:

Martin, A. H., Hoag district, Cal.: Min. Sci., p. 177, 1910.

Preston, E. B., Modoc County: California State Min. Bur. Tenth Ann. Rept., p. 335, 1890.

Stines, N. C., Hoag district, Cal.: Min. and Sci. Press, pp. 384–386, Mar. 13, 1910.

Storms, Wm. H., The High Grade mining district: Min. and Sci. Press, pp. 273–275, Aug. 31, 1912.

TOPOGRAPHY.

The Warner Mountains in the vicinity of High Grade are rugged, but have a fairly even crest at an elevation of 7,500 feet. Mount Bidwell, about 3 miles southeast of the camp, attains an elevation of 8,551 feet.

The west front of the range has an abrupt rise of 1,000 feet above the level floor of Goose Lake valley, which is about 5,000 feet above sea level. This rise is along a fault which follows the east side of the valley for a considerable distance. The scarp is modified by erosion, and several good-sized streams have cut back into the range. The canyon cut by Pine Creek, followed by the High Grade road, has very steep sides. For about 3½ miles eastward from the town of New Pine Creek it has an even rise of about 130 feet to the mile, but in the next 1½ miles there is a rise of 1,000 feet.

The east front of the Warner Mountains was not visited by the writer, but it is shown on the Alturas topographic sheet as an abrupt rise. Russell shows a fault scarp along the west side of Surprise Valley.[1]

---

[1] Russell, I. C., Geological history of Lake Lahontan, a Quaternary lake of northwestern Nevada: U. S. Geol. Survey Mon. 11, 1885.

The mines so far opened in the High Grade district are located along the summit of the range at the headwaters of New Pine Creek, a westward-draining stream, and Deep Creek and Bidwell Creek, which drain northeast and southeast, respectively. The most prominent peak on this divide is Yellow Mountain, which has a barometric elevation of 8,000 feet. This peak has a steep western face, but slopes off at a gentle angle to the east under Mount Bidwell. (See Pls. IV and V.)

GEOLOGY.

General features.—The lower 500 feet of the hills immediately east of Goose Lake is made up of purplish and greenish beds of andesite tuffs, which form the walls of Pine Creek canyon for about 2 miles from its mouth. Overlying these tuffs and dipping east at low angles there is a light buff-colored to white granular rhyolite, which forms the canyon walls to a point about one-fourth mile above the sawmill, where there is a series of dark-brownish soda rhyolite flows, followed by the andesites shown on the left side of Plate V, which are the oldest rocks in the vicinity of the mines.

In the High Grade district proper there are four distinct types of extrusive lavas. The oldest rock exposed along the lower western side of the district (see Pl. V) is a dark fine-grained andesite; above this is a white to yellow rhyolite that in some places is tuffaceous, but in most places is finely granular; above this rhyolite on the flat eastward slope of Alturas Ridge (see Pls. IV and V) there is a purplish flow rhyolite with very fine lamellæ. A fresh porphyritic augite basalt is exposed on the west flanks of Mount Bidwell, and presumably forms the summit of this eminence. Dark obsidian (volcanic glass) covers a small part of the area shown on Plate V as basalt.

Andesite.—The andesite in the western part of the district (see Pl. V) is composed of dark-gray to green flows, dipping east at low angles. All the andesites show distinct flow structure, and some of them are in part glassy. The porphyritic andesite on the south side of Discovery Hill, though the best-preserved rock of this type seen, is much altered. The rather small plagioclase phenocrysts are largely altered to chlorite and some calcite. The groundmass is composed of plagioclase laths and magnetite, the plagioclase being largely altered. All the ferromagnesian minerals are altered beyond recognition. The feldspar crystals in the groundmass distinctly bend around the phenocrysts.

Contact of andesite with rhyolite.—On the ridge about 2 miles west-northwest of High Grade the yellowish rhyolite overlies the

PANORAMA OF WARNER MOUNTAINS NEAR HIGH GRADE.

Looking south-southeast. Shows level summit line and principal peaks.

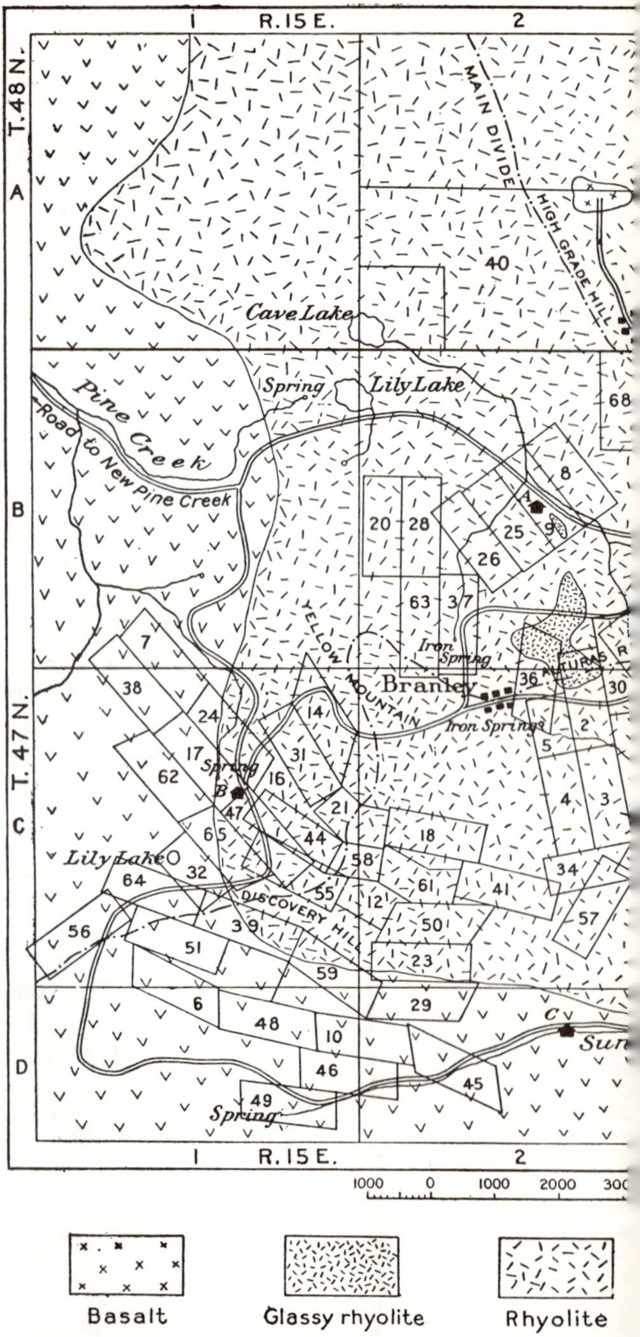

SKETCH MAP OF HIGH GRADE

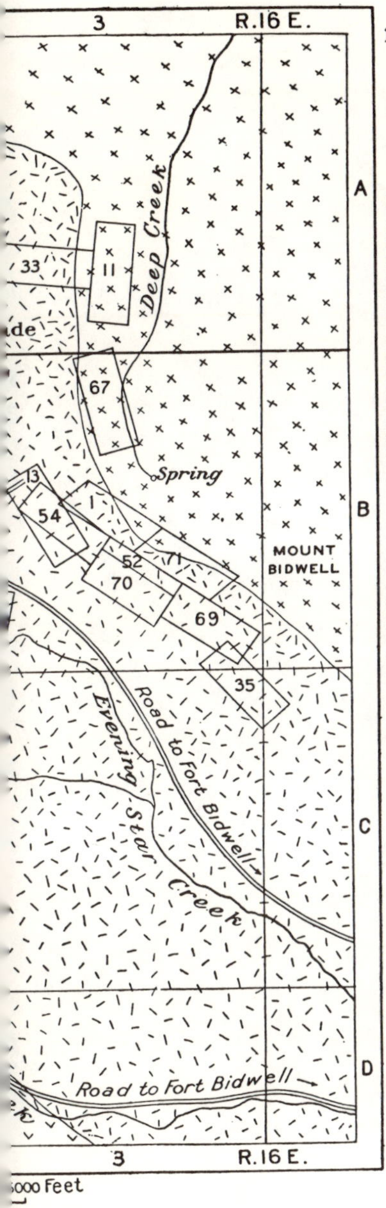

MINES

Names of claims	Coordinates
1. Alaska Bell	B-3
2. Alturas	C-2
3. Alturas No. 1	C-2
4. Alturas No. 2	C-2
5. Alturas Fraction	C-2
6. Archerton	D-1
7. Bad Bill	C-1
8. Big Bonanza	B-2
9. Bonanza King	B-2
10. California	D-2
11. California Bell	A-3
12. Cliff	C-2
13. Comstock	B-3
14. Crown Point	C-1
15. Dandy	C-2
16. Diamond Fraction	C-1
17. Emerald	C-1
18. Eugene	C-2
19. Evening Star	B-3
20. Friday Morning	B-2
21. Gold Peak	C-1
22. Gold Wedge	D-1
23. Happy Thought	C-2
24. Hildebrand	C-1
25. Huckleberry	B-2
26. Huckleberry No. 1	B-2
27. Independence	D-3
28. Josephine	B-2
29. Jumbo	D-2
30. Klondike	C-2
31. Klondike	C-1
32. Leland	C-1
33. Lily Bell	A-3
34. Lonaconing	C-2
35. Mammoth	C-3
36. Mineral Spring	C-2
37. Mineral Springs	B-2
38. Mispah	C-1
39. Modoc Independence	C-1
40. Modoc Mines Co.	A-2
41. Monarch	C-2
42. Morning Star	A-3
43. Mountain Sheep	B-2
44. Mountain View	C-1
45. Nine-Thirty	D-2
46. North Star	D-1
47. Old Glory	C-1
48. Oregon	D-1
49. Peacemaker	D-1
50. Pink Rose	C-2
51. Printer Boy	C-1
52. Red Horse	B-3
53. Red Quartz	B-3
54. Sea Lion	B-3
55. Shasta View	C-1
56. Spearmint	C-1
57. Stonewall	C-2
58. Sugar Pine	C-2
59. Sunset	C-1
60. Sunshine	B-3
61. Tamarack	C-2
62. Topaz	C-1
63. Two Jacks	B-2
64. Uncle Joe	C-1
65. Valley View	C-1
66. White Quartz	B-2
67. White Ribbon	B-3
68. Yellow Jacket	B-2
69. Yukon No. 1	B-3
70. Yukon No. 2	B-3
71. Yukon No. 3	B-3

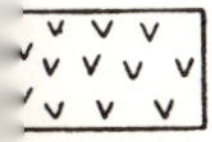
ndesite

Mill

A. Big Four
B. Consolidated
C. Custom

DISTRICT, MODOC COUNTY, CAL.

andesite. In the valley of Pine Creek the contact is covered by glacial material, but it is probable that it is about at the 6,600-foot contour. The contact is again seen about 400 feet below the summit of Discovery Hill, at an elevation of 7,700 feet, but in Sunset Creek it lies at an elevation of only 6,400 feet. The great difference in elevation of the andesite-rhyolite contact at different places is probably due in the main to faulting, though some of it may possibly be attributed to erosion of the andesite surface prior to the outflow of the rhyolite.

Rhyolite.—The rhyolite that forms all of Yellow Mountain and High Grade Hill and the northern part of Discovery Hill is white but weathers yellow. It consists largely of fine granular quartz and orthoclase, with here and there a very small amount of biotite. On the western slopes of High Grade Hill and Yellow Mountain there is about 200 feet of yellow rhyolite tuff below the granular rock. The majority of the rich veins of the camp are found in this rhyolite.

Glassy rhyolite.—On Alturas Ridge (see Pl. V) there is a small area of thinly laminated purple rhyolite. Under the microscope this rock is seen to be in large part a glass in which there are a few crystals of quartz and orthoclase. Its relation to the granular rhyolite described above was not definitely determined, though it appears to be a down-faulted block and is presumably younger than the granular rhyolite.

Basalt.—The augite basalt on the west flank of Mount Bidwell, shown in the upper right-hand corner of Plate V, is a perfectly fresh, greenish-black porphyry that weathers dull brown. The few andesine phenocrysts are quite large and well developed. The groundmass is composed of andesine, pinkish-gray augite, almost colorless olivine, and magnetite, named in the order of decreasing abundance.

On the road from High Grade to Branley, about one-eighth of a mile south of the Pine Creek-Fort Bidwell road crossing, there is a small exposure of black porphyritic volcanic glass, obsidian, which shows columnar structure. On Plate V this rock is mapped with the basalt, though it may have the composition of a rhyolite.

Structure.—The various flows described above have a low eastward dip, being part of the tilted fault block that forms the Warner Mountains. The faults along which this block has been raised are on the west side of Surprise Valley and the east side of Goose Lake valley. Along the Goose Lake valley fault the vertical movement was relatively greater than along the Surprise Valley fault, which resulted in the monoclinal structure of the region visited. There is some evidence of minor faulting near the summit of the

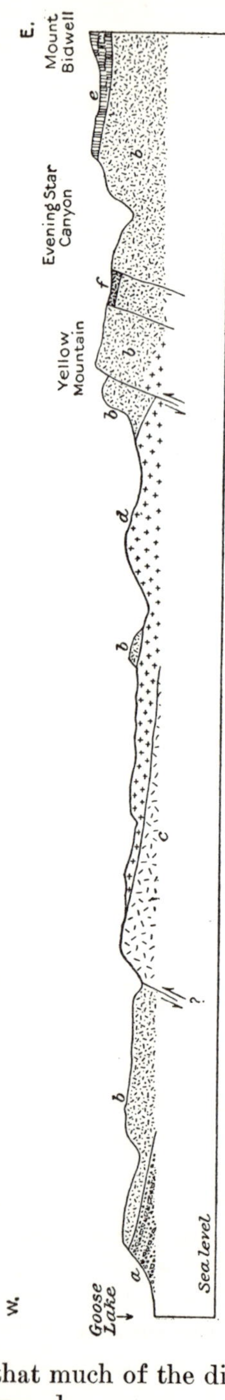

FIGURE 1.—Generalized structure section across the Warner Mountains near High Grade. *a*, Andesite tuff; *b*, granular rhyolite and tuff; *c*, soda rhyolite; *d*, andesite; *e*, basalt; *f*, glassy rhyolite.

range and it seems probable that there is a rather large fault about 3 miles east of the Goose Lake fracture, indicated in figure 1. No definite evidence of this was seen, as the writer spent most of the available time at the mines. There is, however, a strong suggestion of such a fault in the abrupt rise in Pine Creek canyon at that point and a series of cliffs in the south branch of Pine Creek.

The mineralization of the High Grade district took place along the minor fractures near the summit of the range.

"ROCK PILES."

The so-called "rock piles" of High Grade Hill and the north slope of Discovery Hill are, as the name implies, piles of angular talus blocks up to 6 feet in longest dimension. A typical example of these piles is shown in Plate VI, *A*, a view of the Sunshine tunnel. This talus in some places is 20 feet thick but is usually 5 to 10 feet. The principal veins so far disclosed are near these piles, and prospecting for them is rendered difficult on account of the necessity of moving so much loose material. These accumulations of talus are the result of differential weathering. The rock of which they are composed is highly silicified rhyolite, which weathers much less rapidly than the ordinary unaltered rhyolite. This unequal speed of weathering has resulted in the disintegration of the surrounding softer rhyolite, leaving knobs of the more resistant rock. The action of the freezing of water in the small cracks penetrating the rock has split off the large angular blocks.

The depressions around the base of some of the piles are to be explained by the fact that much of the disruption of the knobs has taken place when there was deep snow on which the blocks have fallen. When this snow melted it dropped its load in an irregular fashion. In most of the talus piles some unmelted snow can be found throughout the year.

EVIDENCES OF GLACIATION.

There was a small glacier in the main fork of Pine Creek between High Grade Hill and Yellow Mountain (see Pl. V), which carved the steep north slope on Yellow Mountain and formed the low moraines which have dammed Cave and Lily lakes. (See Pl. IV.) The ice did not move far, so that practically all the material in the moraine is angular. There was also a much smaller ice field in the south head of Pine Creek west of Discovery Hill, which formed the basin now occupied by south Lily Lake. (See Pl. V.)

The glaciation seems to have been restricted to the western side of the Warner Mountains in this vicinity, as there is no evidence of ice erosion in the valleys of either Sunset or Evening Star creeks.

ORE DEPOSITS.

TYPES.

There are three rather distinct types of ore deposits in the High Grade district—(1) veins in granular rhyolite, with some replacement of the walls; (2) veins in andesite; (3) veins and replacements in glassy rhyolite. The veins of the kind first named yield the largest amounts of ore. Gold alloyed with some silver is the valuable constituent of the ore and in all places so far opened the gold occurs free. Minor amounts of copper silicate and a little carbonate were seen in the veins in andesite.

ORIGIN AND GENERAL FEATURES OF THE VEINS.

The development work in the district consists largely of open cuts and shallow shafts, so it is not possible to draw final conclusions. It appears, however, that the mineralization of all the types occurred at about the same time. The solutions came up along fractures partly filled with breccia, especially in the rhyolite areas. The principal veins occur in nearly east-west vertical fissures, though north-south and northwest-southeast fractures are not unknown. The mineralization was accomplished by hot siliceous solutions carrying gold, silver, iron, and potassium, which altered the wall rocks near the fissures by the addition of potassium to form quartz and adularia rock. In practically all the veins postmineral movements have in places crushed the ore and produced gouge, which, though seen in places between the wall and the vein quartz, is commonly found in the quartz where the wall rock was too tough to fracture. The clay gouge of almost every vein constitutes the richest ore, showing a secondary downward concentration of gold from the parts of the veins that have been removed by erosion.

VEINS IN GRANULAR RHYOLITE.

Occurrence and development.—The veins in granular rhyolite are found on High Grade Hill, on Alturas Ridge, and the north side of Discovery Hill. In all these places the granular rhyolite has been more or less silicified, particularly along zones of brecciation, the result being that the rock has become exceedingly hard. On High Grade and Discovery hills this hard, altered rhyolite has weathered into the so-called "rock piles." (See Pl. VI, *A*.) The veins of this type are usually associated with these areas, though there are some veins in unbrecciated rhyolite. In these veins the best ore occurs in shoots which pitch east or southeast on the veins at medium angles.

The Yellow Jacket vein (No. 68, Pl. V) shows most clearly the character of ore from the brecciated silicified zones. In this fracture there are 10 inches to 2 feet of breccia, partly cemented by quartz and some adularia, with small open druses lined with clear quartz needles, as shown in Plate VII. The light-colored bands near the edges of the rhyolite fragments contain adularia and a little pyrite. The dark bands that come next are stained by iron oxide; outside of this the first materials deposited in the cavities are irregular light-colored bands, which are generally quartz but in some places show crystals of adularia as well; clear white quartz was the last mineral deposited in the process of crustification. The gold occurs as very small particles in the dark and light colored bands between the altered rhyolite and the white comb quartz. It is rather light yellow and apparently contains considerable silver.

The Sunshine vein cuts silicified rhyolite breccia, but the ore consists of milky-white to gray quartz and adularia which has been crushed and later clear glassy quartz needles deposited in the breccia.

The Mountain View vein was not studied, but it is said that the rhyolite it cuts is not brecciated, though the ore contains brecciated wall rock with quartz.

On the cliffs between the two heads of Evening Star Creek there are several prospects in granular rhyolite, some of which show no veins. The rhyolite is iron stained and is said to assay fairly well in gold over a width of several feet. These areas are all near fissures, and this mineralization may possibly be somewhat similar to that seen in the walls of the Modoc vein, which are said to be gold-bearing for several feet from the fissure.

A. SUNSHINE TUNNEL.

Looking S. 51° W. Shows typical "rock piles" of High Grade Hill.

B. CROPPINGS OF THE OREGON (KAFADER) VEIN.

Looking N. 65° W.

BRECCIATED ORE FROM YELLOW JACKET CLAIM, HIGH GRADE DISTRICT.

The development of this type of deposits is more extensive than that of any of the others, but it has been scarcely sufficient to show more than the general features of the veins. A few of the more important workings are described below.

Modoc mines.—The Modoc Mines Co. owns the south half of sec. 36, T. 48 N., R. 15 E. (see Pl. V, p. 40) with the exception of the Last Dollar claim in the southwest quarter of the section. The country rock under all this ground is granular rhyolite, with the exception of a few small areas of obsidian shown on Plate V. The crest of High Grade Hill is a "rock pile."

In the early part of August, 1912, the principal development was a 60-foot shaft with short drifts at the 50-foot level, in the northeastern part of the ground, though there were several pits and open cuts near the center of the property on High Grade Hill.

The main shaft is sunk on a nearly vertical fracture that strikes N. 69° E. in silicified fault breccia. The fissure is 3 to 15 inches in width, and has heavy auriferous clay gouge on both walls. The center of the vein carries 1 inch to 4 inches of beautifully banded, milky-white and dark-gray quartz and adularia of unknown value. The highest-grade ore is a soft, heavily iron-stained breccia of wall rock, quartz, and clay between the central quartz band and the walls of the vein. In this ore the matrix of the breccia has been partly leached, whereas the original fragments have suffered less alteration. The wall rock is somewhat iron stained for several feet from the vein, and numerous small quartz stringers run through it in all directions. A little pyrite altering to limonite was seen in some wall rock on the dump that was said to have come from the bottom of the shaft.

Yellow Jacket.—The north end of the Yellow Jacket claim (No. 68, Pl. V, p. 40) is underlain by the silicified rhyolite breccia, in which there are a number of fractures that strike about N. 75° W. One of these fractures has been opened by shallow cuts for about 400 feet and seems to continue eastward into the Sunshine ground. The ore consists of very hard silicified breccia described on page 44 and illustrated in Plate VII.

Sunshine vein.—The Sunshine vein strikes N. 75° W., dips 85° S., and is 2 to 8 inches wide. The development consists of a 350-foot drift tunnel, which attains a depth of about 60 feet and from which there are several small stopes and an air connection. A narrow open fissure that strikes N. 25° E. about 185 feet from the mouth of the tunnel crosses the main vein, but both are apparently of the same age. On the surface the vein has been traced for about 500 feet. It cuts silicified rhyolite that shows only slight brecciation. The richest ore so far taken from this vein is a mixture of a brownish-colored clay gouge and quartz. In some places this clay is 4 inches thick on the walls, and it is also found in the drusy cavities of the siliceous ore.

Mountain View.—The Fort Bidwell Consolidated Mining Co. owns a group of 16 claims on the north side of Discovery Hill. Owing to litigation the main development on the Mountain View (No. 44, Pl. V) could not be studied, though to judge from the dump the ore and country rock are quite similar to those of High Grade Hill, but the vein apparently cuts both rhyolite and the underlying andesite. It is said that in the main tunnel the first 150 feet are crosscut, and that beyond the turn the drift follows the vein for 175 feet. From this tunnel there are said to be two winzes. There was in 1912 a 10-stamp straight amalgamation mill at the tunnel mouth.

Sugar Pine.—The development on the company's Sugar Pine claim (No. 58, Pl. V) consists of three drift tunnels on a vein that strikes N. 80° W. and dips 60°–70° S. The vein ranges in width from 4 inches to 5 feet and consists of partly cemented brecciated rhyolite. In some places the fragments are coated with small comb quartz needles and in others the cementing material is granular quartz. Throughout all the vein there is considerable clay that carries gold, and it is said that the wall rock for 2 feet on either side of the vein is a low-grade ore.

Big Four.—The vein at the Big Four shaft (No. 9, Pl. V) strikes N. 30° W. and dips 70° NE. It is developed by a 100-foot inclined shaft and drift tunnel, from which there has been some stoping. It ranges from a tight fracture to a vein $2\frac{1}{2}$ feet wide and cuts granular rhyolite. The ore occurs in a 150-foot shoot in the wider part of the vein, pitching southeast at steep angles. It consists of fragments of rhyolite and quartz in a clay matrix and is said to yield an average of about $25 a ton.

VEINS IN ANDESITE.

Occurrence.—The veins in andesite in the area visited are confined to the south side of Discovery Hill. The andesite is cut by a number of small white quartz stringers, which strike west-northwest and east-northeast and which in some places unite to form fair-sized veins, as on the North Star, Oregon, and Sunset claims. Most of the larger veins strike about N. 60° W. and dip 70°–75° S. Post-mineral movement has formed gouge along the walls and in some places has crushed the veins.

Oregon.—The original discovery of the district was made on the Oregon claim (No. 48, Pl. V). The vein ranges in width from about 4 inches to 1 foot. It consists of white quartz with very small quantities of adularia, and in places is stained with thin films of limonite or more rarely bluish-green copper silicate. It stands out very distinctly against the dark andesite walls. (See Pl. VI, *B*, p. 44.) In a few places very small crystals of pyrite were seen, but they are rather scarce. The gold is very fine and can not be seen in the ore, but small colors can be panned. The claim is developed by open cuts and a crosscut tunnel 150 feet long, with about 120 feet of drifting on the main vein. A small subparallel vein is disclosed in a crosscut 18 feet south of the Oregon, but does not appear to be very strong.

Sunset.—The Sunset vein (No. 59, Pl. V, p. 40) cuts both the granular rhyolite and underlying andesite. It is developed on the surface in rhyolite by a series of open cuts, and a long tunnel lower down on the hill in andesite has probably reached the same vein. In the rhyolite it contains the typical brecciated quartz-cemented ore, but in the andesite the vein consists of white quartz, clearly distinct from the walls.

North Star.—On the North Star claim (No. 46, Pl. V) there are two narrow quartz veins developed by a 250-foot drift tunnel and a 70-foot crosscut 600 feet west of the mouth of the drift tunnel.

VEINS AND REPLACEMENTS IN GLASSY RHYOLITE.

At the Alturas shafts (No. 2, Pl. V, p. 40) a 105-foot vertical shaft is sunk along a north-south fracture cutting glassy flow rhyolite. In some places along this zone there is a little fault breccia, but the fissure is usually quite tight. Pyrite has been deposited as the matrix of the breccia, together with bluish, white, and amethystine quartz, and has also followed the crystalline lamellæ of the glassy rhyolite fragments in the breccia as well as in the walls. Pyrite in thin films apparently replaces certain lamellæ of the flow rhyolite for 18 feet east of the fissure. This class of ore is said to run about $5 a ton. The value is in gold and the ore above the 50-foot level is said to have been free milling.

On the Dandy (No. 15, Pl. V) and Bonanza King claims (No. 9, Pl. V) there are bodies of much altered iron-stained glassy rhyolite which are said to carry free gold. The Bonanza King ore is reported to run $25 a ton.

FUTURE OF THE DISTRICT.

There can be no question that some of the veins in the High Grade district are superficially rich. All are narrow, as far as known. There are some bodies of replacement ore in the glassy rhyolite, but they are small, and to judge from the Alturas mine the original ore of this class is not of high grade. Pyrite is the only sulphide seen in the region, and it seems probable that the ore below water level, which can not lie very deep, will prove to be relatively low-grade sulphide ore, such as is impracticable to work in many camps. It seems possible, however, that some of the veins will produce considerable gold from ore above the ground-water level.

WINTERS DISTRICT, MODOC COUNTY, CAL.

LOCATION AND ACCESSIBILITY.

The Winters mining district, including Hess Camp (No. 3, Pl. I, p. 18) is about 35 miles west-southwest of Alturas, 16 miles north of Adin, and 3½ miles south of Pit River. It is shown in the west-central part of the Alturas topographic sheet, lying on the southwest side of the Scheffer Mountain just east of the Adin-Alturas road. The camp can be reached by triweekly stage from Alturas, the county seat, which is on the narrow-gage Nevada-California-Oregon Railway, running between Reno, Nev., and Lakeside, Oreg. The main road west of Alturas follows Warm Spring Valley to a point almost due north of Adin, then ascends a low divide west of Scheffer Mountain, and crosses into the north fork of Ash Creek, on which Adin is located. The mines are on the divide north of Ash Creek and lie chiefly in the Pit River drainage basin.

HISTORY AND PRODUCTION.

Claims were staked about 1902, but it was not until 1904 that T. C. Hess located the principal vein in the district. There has been only mild local excitement over these deposits, and though a few men are doing assessment work on several claims there is but one producing mine. Since 1910 the Hess property has been leased to P. H. Keegel and J. L. Harvey, who have erected a 20-ton cyanide mill on Pit River, 3 miles north of the mine. The mill is equipped with ten 500-pound stamps, two 4 by 10 foot amalgamation plates, and five 17-ton cyanide leaching tanks. In the summer of 1912 a 6-ton slime agitator was being installed. The ore fed to this mill is said to carry $12 to $15 a ton in gold and silver, the ratio of gold to silver being about 8 to 1. The production during the past two years is said to have been about $1,000 to $1,500 a month.

ECONOMIC CONDITIONS.

Excellent pine timber is plentiful in this region. Water is not abundant at the mine, but small springs are numerous. The winters are said to be long and severe, and heavy snow at times renders outside work impossible.

GEOLOGY.

This district lies in the great area of volcanic rocks which covers so much of northeastern California and southwestern Oregon. The papers listed below contain descriptions of these lava flows:

Diller, J. S., Notes on the geology of northern California: U. S. Geol. Survey Bull. 33, 1886; Geology of the Lassen Peak district: U. S. Geol. Survey Eighth Ann. Rept., pt. 2, pp. 395–432, 1889; Geology of the Taylorsville region, Cal.: U. S. Geol. Survey Bull. 353, 1908.

Diller, J. S., and Patton, H. B., The geology and petrography of Crater Lake National Park: U. S. Geol. Survey Prof. Paper 3, 1902.

Hague, Arnold, and Iddings, J. P., Notes on the volcanoes of northern California, Oregon, and Washington: Am. Jour. Sci., 3d ser., vol. 26, pp. 222–235, 1883.

Le Conte, Joseph, On the great lava flood of the West and on the structure and age of the Cascade Mountains: Am. Jour. Sci., 3d ser., vol. 7, pp. 167–180, 259–367, 1874.

Marcou, Jules, Exploration for railroad route near the thirty-fifth parallel: 33d Cong., 2d sess., H. Ex. Doc. 91, 1856.

Newberry, J. S., Remarks on some lava flows and lacustrine deposits of Idaho and Oregon: New York Acad. Sci. Trans., vol. 1, pp. 53–56, 1882.

Russell, I. C., Preliminary report on the geology and water resources of central Oregon: U. S. Geol. Survey Bull. 252, 1905; Notes on the geology of southwestern Idaho and southeastern Oregon: U. S. Geol. Survey Bull. 217, 1903.

The rocks in the vicinity of the mines are all porphyritic and, with the exception of the cap rock on Scheffer Mountain and that south of the Adin-Alturas road, were thought in the field to be andesites. The cap rock is a vesicular basalt. The supposed andesite porphyries at the south base of Scheffer Mountain are gray to brownish gray and are all more or less altered, though the plagioclase phenocrysts are still prominent. Some of the rocks are thin flows, some are distinctly tuffaceous, and a large proportion are rather massive porphyries.

Under the microscope these rocks are seen to be very similar in composition. The groundmass is usually a mosaic of microscopic crystals of basic plagioclase with magnetite and a ferromagnesian mineral that seems to be augite. In this groundmass, which in all the slides constitutes over half of the mass of the rock, are set small, well-developed crystals of olivine and of basic plagioclase, whose wide extinction angles show it to have nearly the composition of bytownite or labradorite. There are also some rather large irregular granules of magnetite in the rock. The olivines are all partly altered to calcite and serpentine, and in some slides iddingsite is seen in the cracks of the olivines. The feldspars have been largely changed to chlorite and calcite, and the groundmass in all the slides is more or less serpentinized.

ORE DEPOSITS.

General features.—The ore deposits strike west-northwest in a zone of slight displacement along which the flows north of the fault zone have been raised with relation to those south of it. This zone has been traced for about 2 miles east and west and is scarcely more than one-half mile wide. There are in all about 30 claims on which more or less work has been done. Those at the eastern end of the zone are owned by Mr. W. H. Winters, for whom the district is named. Those at the west, with the exception of some claims held by Winters, and the Hess mine, are controlled by the Dixie Queen Mining Co.

As there is very little relief along the ore zone the development will of necessity be largely by shafts. There are a few short tunnels, mostly crosscuts in barren rock, and a number of shafts from 20 to 50 feet deep on the veins, but the greater part of the work so far done consists of prospect pits and cuts. The veins exposed are usually small; the widest seen is the Hess, which attains a maximum width of 8 feet.

Character of the veins.—Brecciated country rock is associated with all the ore mined, but at the Hess mine it is not abundant. Sulphide ore was seen only on the 120-foot level of the Hess mine, where a very small amount of pyrite was found.

The vein filling consists of calcite, largely replaced by quartz and adularia. This replacement process is well shown in a thin section of ore from the 100-foot level of the Hess mine, in which veinlets of quartz and adularia are seen penetrating calcite along cleavage lines. (See Pl. VIII.)

In a thin section of ore from the Dixie Queen claim the same process is shown and also a later development of long needles of gypsum penetrating all three of the original constituents of the ore.

Origin of the veins.—These veins were deposited by solutions rich in compounds of calcium and potassium and in silica. The calcium was deposited as calcite first. Later, but probably during the same period of mineralization, solutions that were rich in silica and potassium changed the calcite into quartz and adularia. It is not known when the gold was deposited.

Age of the deposits.—Beyond the fact that the veins were formed after the brecciation of Tertiary lavas there is no direct evidence as to the age of these veins, though they do not penetrate the last basaltic flows, which are possibly of Pleistocene age. However, they are probably to be correlated with "the late Tertiary epoch" of mineralization of Lindgren.[1]

[1] Lindgren, Waldemar, Metallogenetic epochs: Canadian Min. Inst. Jour., vol. 12, pp. 102–113, 1909.

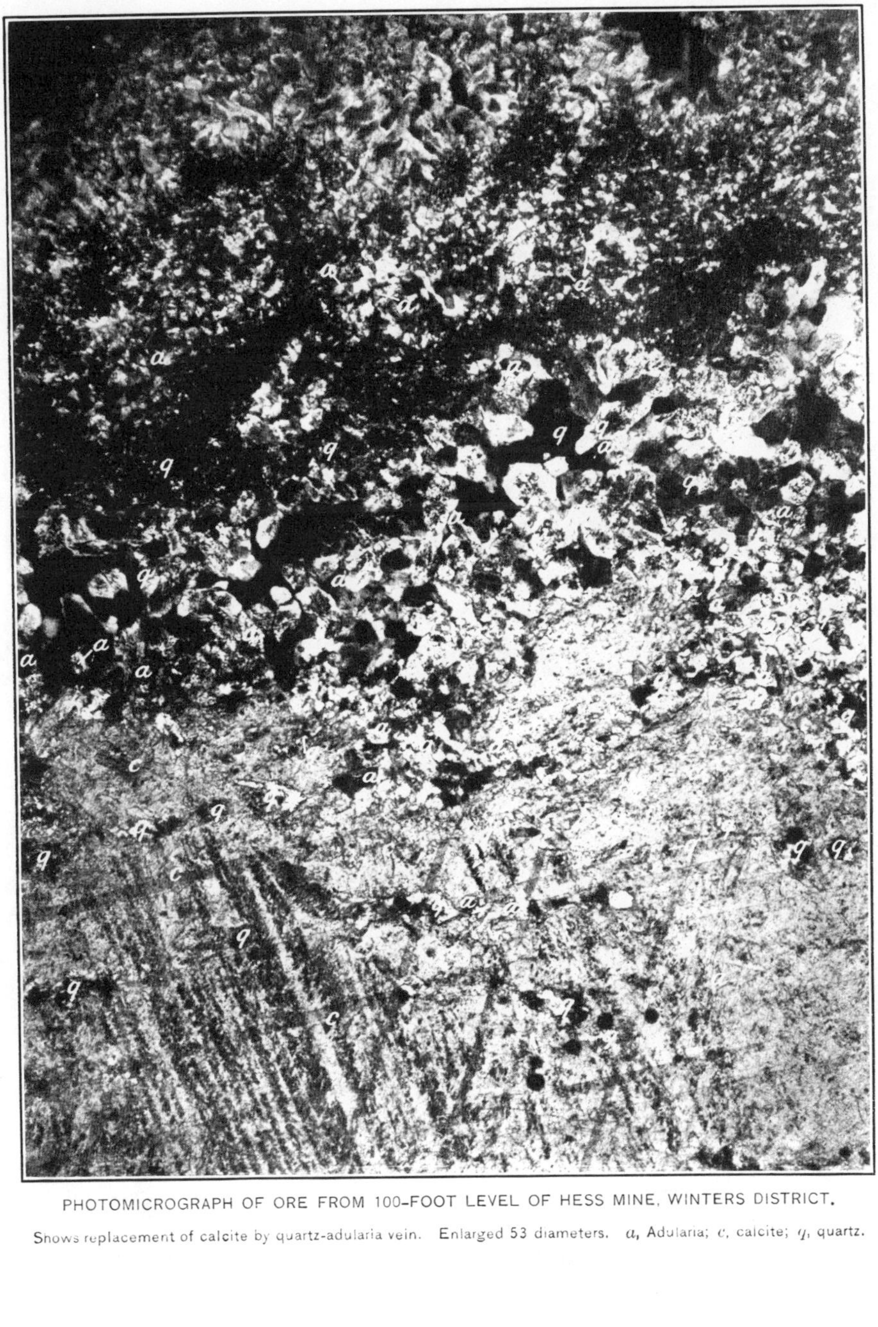

PHOTOMICROGRAPH OF ORE FROM 100-FOOT LEVEL OF HESS MINE, WINTERS DISTRICT.

Shows replacement of calcite by quartz-adularia vein. Enlarged 53 diameters. *a*, Adularia; *c*, calcite; *q*, quartz.

U. S. GEOLOGICAL SURVEY
EORGE OTIS SMITH, DIRECTOR

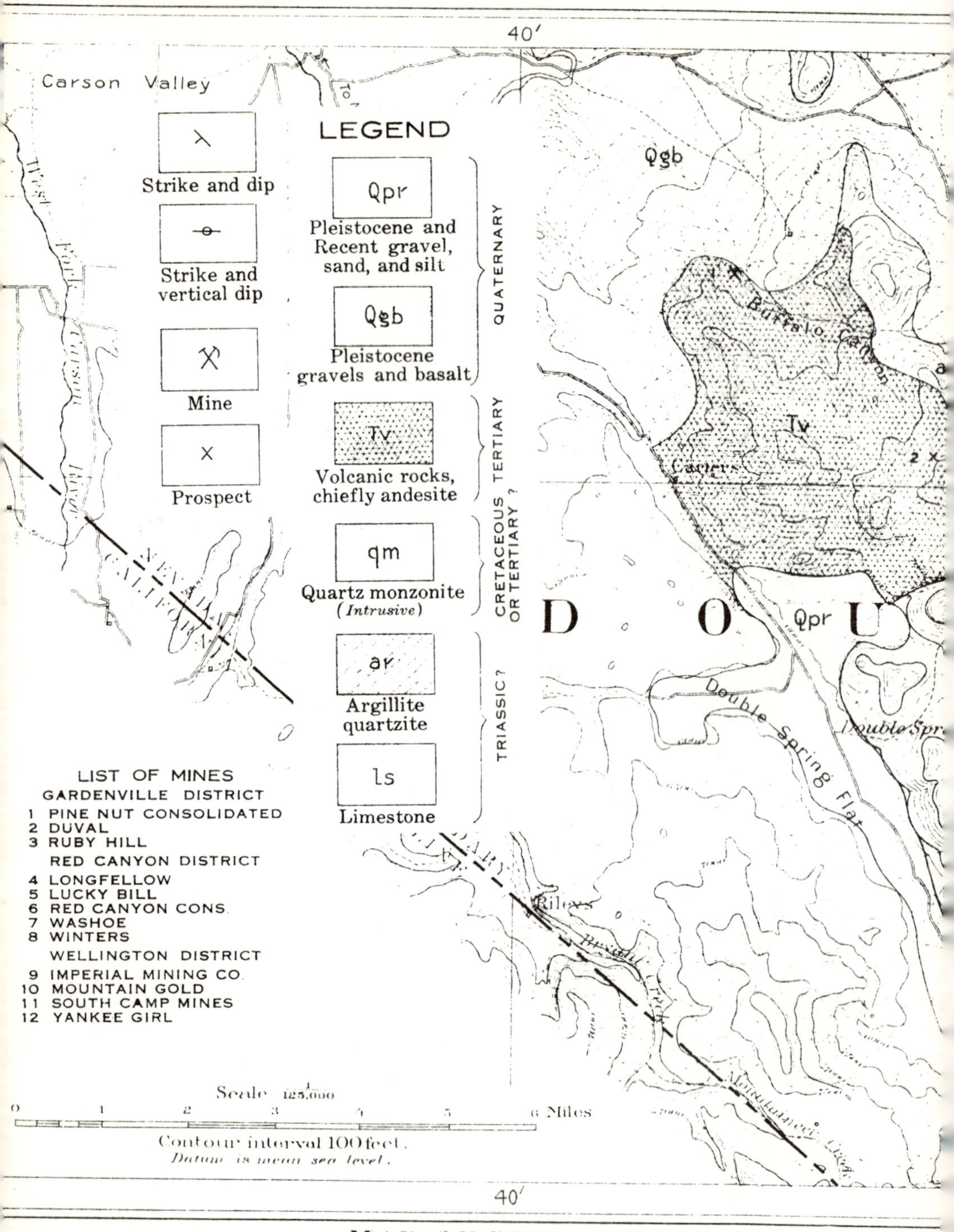

BASE FROM TOPOGRAPHIC MAPS OF
LEEVILLE AND WELLINGTON QUADRANGLES

MAP OF THE SOUTH END OF THE P

Showing approximate boundaries
prospects in the Gardenvill

RANGE, DOUGLAS COUNTY, NEVADA

mations and location of the principal

nyon, and Wellington districts

Hess mine.—The only mine that has been developed to any extent, and the one producer of the district, is the Hess, located about one-fourth of a mile east of the Adin-Alturas road, at the head of the south fork of Stone Coal Creek.

The Hess vein strikes east and dips about 55° S. It occurs along a zone of faulting, which has produced a breccia of basalt 4 to 10 feet wide. In this breccia the vein is 4 to 6 feet wide, consisting of quartz, calcite, and adularia cementing fragments of the wall rock, the latter constituting less than 10 per cent of the ore. This body at the surface is about 80 feet long, at the 120-foot level it is 150 feet long, and at the 150-foot level it is 65 feet long. The inclined shaft is 200 feet deep, at which level the vein, which was 6 feet thick at the 150-foot level, has split into a great number of small, slightly mineralized quartz stringers extending through 20 feet of soft gougelike crushed basalt.

The gold, which occurs in very fine particles, is said to be hard to save by amalgamation, but to be easily extracted by the cyanide process. The best ore is said to occur near the footwall, but at the 100-foot level 3 to 4½ feet of good ore was found above the footwall under about 1 foot of lower-grade ore.

The ore body exposed in the shaft has been at least three-fourths stoped, largely by overhand methods. The stopes are well timbered but open. As yet no new ore has been developed, though it was the intention to prospect the vein east of the main shoot and below the present development.

Six men are employed at the mine, working two shifts; the hoisting is done by a whim capable of working to a depth of 500 feet.

COAL.

Stone Coal Creek was named for a deposit of rather impure coal found in an eastern branch north of Hess camp and about 2 miles south of Pit River. The deposit was not visited but is reported to be small. It doubtless can be correlated with the coal-bearing Ione (Neocene) formation described by Diller[1] as occurring in Little Cow Creek valley, in the northeastern part of Sacramento Valley.

SOUTH END OF PINE NUT MOUNTAINS, DOUGLAS COUNTY, NEV., INCLUDING GARDNERVILLE, RED CANYON, AND WELLINGTON MINING DISTRICTS.

LOCATION AND ACCESSIBILITY.

The area next to be described lies at the south end of the Pine Nut Mountains, between Gardnerville, in Carson Valley, and Well-

[1] Diller, J. S., U. S. Geol. Survey Geol. Atlas, Lassen Peak folio (No. 15), 1895.

ington, in Smith Valley, east of the Pine Nut Mountains. The central part of this area is at about 38° 50′ north latitude, 119° 30′ west longitude. In this vicinity there are three areas, which in this report are known as the Gardnerville, Red Canyon, and Wellington mining districts, shown at localities 4, 5, and 6 on Plate I (p. 18). This region is shown on the Markleeville and Wellington topographic sheets, from which Plate IX is adapted.

The Wellington district (No. 6, Pl. I) lies in the southeastern part of the range, the mines being largely on the eastern and southern slopes of the mountains, adjacent to the town of Wellington. It can be most easily reached from Hudson station, on the Nevada Copper Belt Railroad, which is on the east side of Smith Valley 10½ miles northeast of Wellington. The mines in the Red Canyon district (No. 5, Pl. I) are in Red Canyon and on the crest of the mountains at the head of that canyon. Hudson is 6 miles almost due east of the mouth of the canyon, but the mines on the summit are more easily reached from Minden and Gardnerville, in Carson Valley, west of the mountains. Minden, 1 mile northwest of Gardnerville, is the terminus of the south branch of the Virginia & Truckee Railway.

The Gardnerville district (No. 4, Pl. I), as discussed in this report, covers the low foothills at the southeast end of Carson Valley and is largely drained by Buffalo Canyon. (See Pl. IX.) Daily stages operate between Gardnerville and Wellington and between Hudson and Wellington.

In Antelope, Carson, and Smith valleys a large part of the land is under irrigation. As Carson and Smith valleys are served by good railroads and are favorably situated, large crops of alfalfa and grains are annually shipped from them to towns in this region.

HISTORY AND PRODUCTION.

As far as can be learned the first ore mined in the south end of the Pine Nut Mountains was taken from the Longfellow vein (No. 4, Pl. IX) about the year 1862 and was hauled to Virginia City for treatment. The Winters ore body was discovered in 1872 and for a time rich surface ores were taken out. The other mines in these three districts have been exploited in recent years, the latest discovery being the Ruby Hill (No. 3, Pl. IX), which was located in 1908 by Mr. Walker, the present superintendent. The Yankee Girl mine on Taylor Hill, about 3 miles northwest of Wellington, was worked to some extent in the eighties, the ore being treated in arrastres on West Walker River, just west of Wellington. In the summer of 1912 the Ruby Hill mines were shipping ore and development was under way at the Winters, Longfellow, and Imperial mines and the Mountain Gold and Lucky Bill groups.

There are two cyanide mills in this region. The Winters mill, located near the mine, has three 1,000-pound stamps and four wooden cyanide leaching tanks, capable of treating 25 tons of ore daily. At the Pine Nut Consolidated property the mill is equipped with crusher, rolls, amalgamation plates, Frue vanners, and two 50-ton leaching tanks for treating the slimes.

There were two small amalgamation mills in the district, but these have not been used for several years. The Longfellow mill, built about 1872, $1\frac{1}{2}$ miles west of the shaft, where water is abundant, had five 100-pound stamps and a short amalgamation plate. There was formerly a small 2-stamp mill at the South Camp property.

There are few adequate figures of early production from the mines in this region. It is estimated that about $50,000 was taken from small rich pockets in the Longfellow vein. The Taylor mine has undoubtedly produced some ore, and the Winters mine has shipped several thousand dollars worth of ore and bullion.

The figures of production for the Red Canyon and Gardnerville areas in the years 1903 to 1911, inclusive, published by the United States Geological Survey, are given in the following table, though it is not certain that the figures for Gardnerville do not include the output of some of the mines on the summit of the range which in this report are placed in the Red Canyon district. No report of production has been made for these years from mines in the Wellington district.

Production of mines near Gardnerville and in Red Canyon district, Douglas County, Nev., 1903 to 1911, inclusive.

Gardnerville.

Year.	Tonnage.	Gold.	Silver.	Copper.	Total value.
			Ounces.	*Pounds.*	
1905	30	....	1,328	35,000	$2,259
1907	57	$274	286	3,593	1,182
1908	271	1,486	21	....	1,497
1910	11	170	16	....	213
1911	38	1,500	134	....	1,900
	407	3,430	1,785	38,593	7,051

Red Canyon district.

Year.	Tonnage.	Gold.	Silver.	Lead.	Total value.
			Ounces.	*Pounds.*	
1903	300	$1,700	4,764	....	$3,700
1904	1,020	1,230	26,321	....	16,233
1905	565	300	2,099	4,000	1,756
1910	12	317	6	....	320
1911	5	....	....	....	105
	1,902	3,547	33,190	4,000	22,114

TOPOGRAPHY.

The south end of the Pine Nut Mountains is a rugged mass, 12 miles in width, which rises to a maximum height of 9,500 feet at Galena Peak and 9,350 feet at Bald Mountain. The east side of the range rises steeply from Smith Valley and is cut by a number of deep, narrow canyons. The largest of these, Red Canyon, was the northernmost visited in this reconnaissance. At the south end the range is separated from the Sweetwater Mountains by the canyon of West Walker River, from which the mountains rise to an elevation of 7,000 feet in 3 miles.

The west side of the range, sloping into the Carson Valley, is much less abrupt than the eastern side. Southwest of Gardnerville there is a large area of low foothill country before the main range is reached.

The low pass (elevation 6,000 feet) from Carters to Mountain House (see Pl. IX) separates the Pine Nut Mountains from the main Sierra. This gap at Mountain House is about 1 mile wide and mountains 7,000 feet above sea level rise within a short distance on either side.

Buffalo and Mill creeks and their tributaries on the west side of the range drain in general northwestward into the Carson Valley. Northward from an east-west line through Bald Mountain there are good perennial springs in all of the canyons, but south of that line there is practically no water in the hills except in the spring of the year.

GEOLOGY.

SEDIMENTARY ROCKS.

Triassic system.—Rocks of supposedly Triassic age are found in the northern part of the mountainous area studied in Red Canyon, on the upper part of the west slope of the range, and in the south fork of Lone Pine Creek. The correlation of these rocks with the Triassic is based largely on the lithology and the negative evidence that no sediments older than Triassic are known in this latitude between the Battle Mountains in central Nevada and the crest of the Sierra Nevada, which lies west of the Pine Nut Mountains. Spurr[1] found Triassic fossils in shaly limestone at Eldorado Canyon, on the northwest side of the Pine Nut Mountains. Ransome[2] regards the limestones in the vicinity of the Ludwig mine, on the northeast side of Smith Valley, as of possibly Triassic age.

[1] Spurr, J. E., Descriptive geology of Nevada south of the fortieth parallel and adjacent portions of California: U. S. Geol. Survey Bull. 208, p. 122, 1903.

[2] Ransome, F. L., The Yerington copper district, Nev.: U. S. Geol. Survey Bull. 380, p. 104, 1909.

A few fossils were found in some dark shale beds near the Lucky Bill mine in Red Canyon, but they are not diagnostic.

The section exposed in Red Canyon is as follows, beginning with the bottom:

Section of Triassic rocks exposed in Red Canyon, Douglas County, Nev.

	Feet.
Thick-bedded white crystalline limestone	100–300
Thin-bedded dark-gray to black argillites with some lenses of white quartzite	800–1,000
Massive blue-gray limestone	1,500–2,000

Southwest of Oreana Peak (see Pl. IX) the argillites are much contorted and metamorphosed, particularly along the contacts with the intrusive granodiorite. The bedding stands nearly vertical in most places, but small tight folding is commonly seen. The limestone in the eastern part of Red Canyon occurs in a tight anticlinal fold whose axis strikes about N. 80° W. The stream has cut across this axis at a very acute angle. Southeast of Wellington a narrow belt of light-colored, tightly folded crystalline limestone is exposed just west of the gravels in Smith Valley. It is highly metamorphosed along intrusive dikes of porphyritic quartz monzonite and the bedding is nearly vertical at all places.

The section in Red Canyon is in some respects similar to the Star Peak Triassic section of the Humboldt Range, described by the geologists of the Fortieth Parallel Survey,[1] who measured the following section, beginning at the bottom:

Section of Triassic rocks at Star Peak, Humboldt Range, Nev.

	Feet.
Limestone	1,500
Slaty quartzite (capped with black slates, 250 feet)	1,500
Heavy ferruginous limestone	2,000
Limestone (owing to peculiarities of structure thickness somewhat in doubt), probably	1,000
Pure quartzite	2,200–2,800

Louderback[2] says: "The Triassic group (of the Humboldt Range) consists chiefly of shale or slate and impure, shaly, gray or drab to black limestones, with here and there a thin layer of quartzite, which may, however, occasionally reach several hundred feet in thickness."

On the west flank of the mountains west of Galena Peak, between 6,500 and 7,500 feet above sea level, thin-bedded dark argillites are exposed, but they are so much contorted and metamorphosed that no estimate of their thickness was made.

---

[1] King, Clarence, Systematic geology: U. S. Geol. Expl. 40th Par., vol. 1, p. 267, 1878.

[2] Louderback, G. D., Basin structure of the Humboldt Range: Geol. Soc. America Bull., vol. 15, p. 205, 1904.

In the Lone Pine Creek drainage basin the area shown on Plate IX as Triassic is underlain by a series of much-altered schistose quartzitic slates and limestones. The white limestones are well exposed in the low hills 2¾ miles southeast of Pete Anderson's ranch, and are the lowest members of the series, as they are overlain by dark schists and light quartzitic schists, the latter predominating. Immediately over the quartzitic schist there is a very thin flow of finely laminated white rhyolite, followed by augite andesite and andesite flow conglomerates. The andesitic material has been mapped on Plate IX as Tertiary, though its age is not certain and it may be similar to the Koipato formation of the Humboldt Range. The basis for mapping this rock with the Tertiary volcanic rocks is the fact that lithologically and mineralogically the andesite resembles some of the andesitic flows of Tertiary age south of Wellington.

IGNEOUS ROCKS.

Intrusive rocks.—A large part of the south end of the Pine Nut Mountains (see Pl. IX) is composed of a granular igneous rock of intermediate chemical composition that is intrusive into the supposed Triassic sediments. The contact of this intrusive rock with the sediments is marked in some places by a zone of metamorphism. A rock of similar composition intruded into the limestones in the Yerington district is said to be late Jurassic or early Cretaceous by Ransome,[1] and intrusions ranging from quartz monzonite to granodiorite of late Cretaceous or early Tertiary age are common along the whole of the Sierra Nevada.

The typical rock of this kind in the Pine Nut Mountains is a gray to pink-gray, coarsely granular to porphyritic variety of quartz monzonite. The contact phases are in some localities fine-grained granular basic rocks, and some small dikes, offshoots from the main mass, are rather fine grained porphyries.

In all the porphyritic facies the chief phenocrystic mineral is pinkish orthoclase. Partly developed crystals of this mineral are set in an unequally granular anhedral matrix whose constituents, named in the order of decreasing abundance, are oligoclase-andesine, orthoclase, quartz, brown biotite, and green hornblende. Magnetite and titanite with a little apatite are sparingly accessory minerals.

On surface alteration this rock changes to a dull pinkish-gray color, owing to the alteration of the plagioclase feldspar to sericite and the partial sericitization of the orthoclase. During this process the ferromagnesian minerals are more or less completely changed to chlorite and in places epidote is developed.

In the quartz monzonite there is a very distinct jointing that strikes west to N. 60° W. and dips 60°–70° S. In some places a less

---

[1] Ransome, F. L., op. cit., p. 104.

distinct north-south vertical jointing is seen, particularly in the mountains southeast of Bald Mountain.

Dike rocks in the quartz monzonite.—In the quartz monzonite area there are a number of nearly vertical dikes that strike north. These dikes as seen in the field are of three distinct kinds. The oldest type is a fine-grained light-colored feldspathic rock, the next younger type is represented by light-gray distinctly porphyritic rocks, and the youngest are very dark, fine-grained porphyries, which at many places cut both the other types.

Thin sections of the light-colored dikes show the rock to be a fine-grained phase of the quartz monzonite, consisting of orthoclase, oligoclase-andesine, quartz, and a few flakes of brown biotite, the numerals being named in order of decreasing abundance. In one thin section the quartz and orthoclase were intergrown, and all the rocks examined under the microscope were altered to a marked extent, the feldspars being changed to sericite and the biotite to chlorite.

The next younger dikes are distinctly porphyritic in the hand specimens, showing phenocrysts of andesine, biotite, and orthoclase; these minerals are well crystallized, the plagioclase showing zonal growth. The groundmass of this dike rock is composed of small interlocking grains of orthoclase, quartz, andesine, biotite, augite, and magnetite, named in order of decreasing abundance. Some epidote is developed along the borders of these dikes, and the mafic minerals are in most thin sections partly altered to chlorite. These dikes range between quartz monzonites and quartz latites in mineralogic composition, and are probably derivatives of the quartz monzonite magma. They closely resemble some of the smaller offshoots from the main mass seen in the sedimentary rocks of Red Canyon.

The fine-grained dark porphyry dikes which cut both types previously described were seen only at the south end of the mountains, though they may occur elsewhere. In this rock the phenocrysts were probably andesine and augite, but in all thin sections they are altered, the andesine to calcite and some sericite and the augite to epidote and chlorite. The groundmass, also altered, is made up of small andesine laths and grains of augite and magnetite. These rocks are classified, without much question, as altered augite andesites. They bear a close resemblance to some of the lower flows of andesite seen in the hills southwest of Wellington that are with little question referred to the Tertiary, but they may be basic differentiations of the quartz monzonite magma.

Contact metamorphism.—The most typical area of contact metamorphism lies at the mouth of Red Canyon, where the massive limestone is cut by a 10-foot dike of quartz monzonite porphyry which has a general northwesterly trend. Along this dike the limestones are altered to dark-greenish rocks that are largely composed of epidote, quartz, and calcite, in which there is some pyrite and chalcopyrite. This zone is about 100 feet in width. Some dark argillite beds in the limestone have suffered much less alteration than the calcareous rock and apparently are less changed than the contact phases of the intrusive, which are altered to brownish-colored, finely granular masses of quartz, orthoclase, sericite, biotite, and brownish amphibole.

At one of the shafts of the Winters group (No. 8, Pl. IX) a lens of impure dark calcareous argillite in quartzite has been changed to a mass of epidote and magnetite, together with some pyrite and chalcopyrite at the contact with quartz monzonite.

South of Wellington, near some prospects just outside of the area shown in the lower right-hand corner of Plate IX, the limestone is intruded by a narrow dike of porphyritic quartz monzonite, along which there is a contact zone about 2 feet wide consisting of epidote, magnetite, and some pyrite, together with minor chalcopyrite.

Extrusive rocks.—There are three areas of Tertiary volcanic rocks within the district described in this section of the report. The largest of these lies south and west of Wellington, though only the northern part of this area was seen.

The Buffalo Creek drainage area, southeast of Gardnerville, lies chiefly in rocks of questionable Tertiary age. The summit of Galena Peak and a small part of the northwest head of Red Canyon are underlain by lavas.

The most widely distributed type of rock of Tertiary age is a fine-grained, nearly black porphyritic andesite, composed of andesine, augite, brown biotite, and accessory magnetite. The chief phenocrystic minerals are andesine and augite, though biotite is found in some places. Southwest of Wellington the flows are fresh, but along the east side of Smith Valley and in the Buffalo Creek area southeast of Gardnerville they have been somewhat altered by hydrothermal action. Above the typical flow andesites in the Buffalo Creek drainage area lie about 200 feet of andesitic flow breccias and conglomerates, which dip to the west at low angles. It is not absolutely certain that these flows in the Buffalo Creek drainage area are of Tertiary age. They may belong with the Triassic sediments, as in the type locality of the Humboldt Range[1] and also at Yerington[2] andesitic material is found in the Triassic section.

---

[1] King, Clarence, Systematic geology; U. S. Geol. Expl. 40th Par., vol. 1, pp. 269–274, 1878. Ransome, F. L., Notes on some mining districts in Humboldt County, Nev.: U. S. Geol. Survey Bull. 414, p. 32, 1909.

[2] Ransome, F. L., The Yerington copper district, Nev.: U. S. Geol. Survey Bull. 380, pp. 101–104, 1909.

The upper flows on the hills south of Wellington are glassy hornblende andesites, the phenocrysts being large, greenish-black hornblendes, altered to epidote, and smaller altered plagioclase laths.

The andesite on Galena Peak, at the summit of the range, is a brown, glassy rock containing rather large crystals of distinctly striated feldspar and some crystals of highly altered augite (?).

QUATERNARY DEPOSITS.

Deposits of partly cemented gravel and sand overlie the Tertiary andesites southeast of Gardnerville, forming the low foothills for about 12 miles east and 10 miles southeast of the town. Due east of the town these deposits rise to an elevation of 6,200 feet and here consist of pebbles of andesite and quartz monzonite. On the Gardnerville-Wellington road the gravels overlie the andesites, the contact being at an elevation of 6,000 feet.

In the canyon of West Walker River, just west of Wellington, basalt flows overlie the gravels of Pleistocene age as noted by Spurr.[1] The gravels at this locality are rather obscure, occurring only within the canyon, whereas the Tertiary volcanic rocks appear to form the major part of the hills both north and south.

Some of the recent wash at the north end of Antelope Valley, shown at the center of the lower part of Plate IX (p. 52), probably overlies gravels of Pleistocene age, particularly in the long arm which extends northeast of the andesite area west of Wellington.

The large, relatively flat floors of Carson, Antelope, and Smith valleys are underlain by fine gravels, sand, and silts of Pleistocene and Recent age, and the surface of these valleys is covered with a very fertile sandy loam.

ORE DEPOSITS.

There are four types of ore deposits in the three districts at the south end of the Pine Nut Mountains. The quartz veins in quartz monzonite were the earliest deposits worked. There are contact-metamorphic deposits, largely copper bearing, at the contact of Triassic (?) sediments and intrusive quartz monzonite, and replacement bodies in the sediments at a distance from known intrusives. The fourth type is represented by the deposits in the supposed Tertiary andesite.

QUARTZ VEINS IN QUARTZ MONZONITE.

General features.—The veins in the quartz monzonite area strike north or east about parallel to the jointing of the rock. The northward-striking veins are in general nearly vertical, but in some places

[1] Spurr, J. E., Descriptive geology of Nevada south of the fortieth parallel and adjacent portions of California: U. S. Geol. Survey Bull. 208, p. 126, 1903.

they dip steeply to the west. The eastward-striking veins dip to the south at high angles. They range from a fraction of an inch to 2 feet in width, averaging from 6 to 10 inches. In many places several nearly parallel, closely spaced, small veins are exposed by a single tunnel. There appear to be only a few veins of this type, and they are as a rule not strongly mineralized. However, small pockets of workable ore occur in most of these veins.

The veins consist of white quartz and minor quantities of pyrite and specular hematite as the chief constituents. In the richer lenses galena and chalcopyrite are associated with the pyrite. At the shallow depths so far reached in veins of this type the constituents are more or less oxidized to a rusty quartz with here and there yellow lead carbonate stains or the blue and greenish copper carbonates.

Longfellow mine.—The Longfellow mine (No. 4, Pl. IX, p. 52) is on a vertical vein in quartz monzonite that strikes N 70° W. It is developed by several shallow shafts and tunnels. Sinking was in progress below 80 feet in the main shaft, which is equipped with a 15-horsepower gasoline hoist. There are a few small northward-striking veins on the claims belonging to the Longfellow group. The oxidized sulphide ore is said to carry about $18 a ton, $3 to $4 of which is in silver and the remainder gold. There was very little ore in sight in any of the mine workings in 1912, as no pockets had been recently opened. In amalgamating it is said that 50 per cent of the gold and silver is free, the remainder being combined with the unoxidized sulphides.

Washoe claims.—The Washoe claims (No. 7, Pl. IX) belong to H. A. Danberg, of Gardnerville. There is very little development on any of the small, slightly mineralized eastward-striking veins.

South Camp mines.—At South Camp (No. 11, Pl. IX) there are three caved tunnels and some shallow shafts on a series of quartz veins that strike N. 5° E. in the quartz monzonite near dikes of quartz monzonite porphyry and andesite. The veins as a rule look quite barren, though in places there are small pockets containing partly oxidized pyrite. Postmineral movement occurred along the veins, forming a narrow gouge on the west side and crushing the quartz to some extent.

Imperial claims.—The Imperial claims (No. 9, Pl. IX, p. 52) are about 6 miles west of Wellington, in the low hills at the north end of Antelope Valley. There are a series of veins in these claims that strike nearly due north. Some of these veins stand almost vertical, though others dip west at medium angles. The deepest development is a 72-foot incline, though there are a number of shallow shafts and pits on other veins. The veins range from 4 inches to 2 feet in width. They are white iron-stained quartz containing rather abundant small specular hematite plates in druses and locally small bunches of sulphides. The sulphide is commonly pyrite, though in some veins galena is the principal ore mineral.

Mountain Gold claims.—The seven claims of the Mountain Gold Mining Co. (No. 10, Pl. IX) are near the crest of the range, 4 miles southeast of Bald Mountain and 8 miles northwest of Wellington, the nearest place at which water is obtainable in any quantity. The veins in this locality strike east and dip north at steep angles. They are typical representatives of the quartz veins in quartz monzonite.

Yankee Girl claims.—The Yankee Girl claims (No. 12, Pl. IX, p. 52) are usually called the Taylor Hill mines. They cover two nearly vertical veins, which strike about N. 50° E. on the north side of this hill. They are opened by shallow shafts and tunnels. Some stoping has been done on the northern vein, which consists of 18 inches of iron-stained quartz frozen to the footwall, but having a thin clay parting on the hanging wall. Few sulphides can be seen in the part of the vein left in this work. At an 80-foot shaft on the south vein a little copper carbonate is associated with the limonite-stained quartz.

CONTACT-METAMORPHIC DEPOSITS.

Red Canyon claims.—The principal contact-metamorphic ore deposit is that belonging to the Douglas Mining & Reduction Co., located at the mouth of Red Canyon (No. 6, Pl. IX, p. 52). At this place limestones that contain some shale beds are intruded by a 10-foot dike of quartz monzonite porphyry, which strikes northwest and dips 50° N. For 100 feet southwest of the contact the limestones are altered to a hard mass of epidote, quartz, and calcite containing pyrite, chalcopyrite, and pyrrhotite in different quantities. The sulphides are largely confined to the metamorphosed limestones, the shales being practically barren. The sulphides are found at the surface, but copper carbonates and sulphur occur in the deepest workings, about 75 feet below the surface. The largest lens of ore lies between the dike and the main zone of contact-metamorphic minerals but consists of the same minerals except that the sulphides form a greater part of the mass.

Iron Boy claim.—On the Iron Boy claim near the Winters mine (No. 8, Pl. IX, p. 52), belonging to the same company, a lens composed of magnetite and pyrite, carrying small amounts of chalcopyrite, is formed in black, somewhat calcareous shales at the contact with coarsely granular, somewhat porphyritic quartz monzonite. The lens is 20 feet wide at the surface and about 150 feet long, but at a depth of 100 feet it is 6 feet wide and dips about 50° W., parallel to the bedding of the shale, which underlies a light-colored quartzite bed.

REPLACEMENT LODES IN TRIASSIC SEDIMENTS.

Lucky Bill group.—The Lucky Bill group of claims (No. 5, Pl. IX, p. 52), in the north fork of Red Canyon, is the principal representative of this class of deposits. A few shallow tunnels and open cuts show the replacements to have taken place in fractured quartzite in which there is considerable disseminated pyrite. The principal fracturing strikes east-southeast, about parallel to the axis of the Red Canyon fold. Small, very pockety deposits of argentiferous galena, quartz, and stibnite, together with minor quantities of pyrite and chalcopyrite, occur at various places along the fracture zones. The galena is somewhat altered to cerusite and the stibnite to yellowish-gray silky fibers of valentinite.

Winters mine.—At the Winters mine (No. 8, Pl. IX, p. 52) the ore body is a veinlike replacement of slightly calcareous argillites. The sediments are in the form of a lens included in quartz monzonite. The main ore body is developed by three tunnels through a vertical distance of about 250 feet and a horizontal distance of nearly 1,000 feet. It has the form of a vein that strikes N. 10° E. and dips 75° W. A fairly continuous quartz stringer that ranges from 6 inches to 4 feet in width carries galena and some stibnite, and pyrite and chalcopyrite are seen here and there in small pockets next to the west wall. Numerous feeders come into the main vein from the footwall side, where the argillites are crushed and impregnated with pyrite for a distance of 130 feet. The ore is completely oxidized to a depth of 150 feet, below which a few bunches of sulphides are seen locally.

In the limestone area south of Wellington there are a few prospects on northerly striking zones in which the limestones are crushed. The mineralization is not strong at any place, but small pockets of sand carbonate and some residual galena are found.

ORE DEPOSITS IN SUPPOSED TERTIARY ANDESITES.

In the augite andesite of the Buffalo Creek drainage area, southeast of Gardnerville, there are several prospects and one mine, the Ruby Hill, which in 1912 was producing copper ores.

Ruby Hill mine.—At the Ruby Hill mine (No. 3, Pl. IX, p. 52) the andesites of supposed Tertiary age are cut by two series of fractures. The older fractures strike about N. 25° E. and dip 45° NW. in a zone 100 to 150 feet wide. There are a large number of small closely spaced slips and the andesite is all crushed and altered in this zone. A later vertical fracture that has a northerly strike seems to cut off the zone at the west. In the fractured altered andesite near the northward-striking fault there are small pockets and lenses of oxidized copper ores. These lenses occur along small north-trending vertical slips. The mine is developed by a 150-foot inclined shaft from which there are short drifts at the 60 and 100 foot levels.

The bodies of ore are mostly small; the largest seen was 15 by 20 feet in cross section and 4 to 5 feet thick. It was about parallel to the flat-dipping north-northeast set of slips.

The ores to the greatest depths reached are all oxidized. They consist of highly altered, crushed andesite, stained with copper carbonates and containing small irregular masses and stringers of cuprite. At the 100-foot level some of the cuprite shows remnants of chalcopyrite and what may be chalcocite. The highly copper-stained andesite containing cuprite is said to carry $2.50 in gold and 2 to 3 ounces of silver a ton, whereas the ore at the 100-foot level, which shows the sulphide, is said to carry $6 in gold and 10 to 11 ounces of silver a ton.

The solutions which brought in this ore have altered the crushed andesite to a marked degree, the result being a rather soft rock, which consists largely of sericite and calcite carrying more or less disseminated pyrite. At some places, particularly along the northward-striking fractures, there are siliceous ribs, but these show that they are also altered andesite. In these bands the sulphides are best preserved.

Duval prospects.—About one-half mile west of the Ruby Hill mine, at the Duval prospects (No. 2, Pl. IX, p. 52), there is a zone of silicified andesite that strikes N. 45° E., in which no sulphides are seen, but which carries gold and resembles the typical croppings of late Tertiary gold deposits.

Pine Nut Consolidated claims.—At the Pine Nut Consolidated (No. 1, Pl. IX, p. 52) the country rock is augite andesite. It is cut by a nearly vertical set of fractures that strike north and by a set of fractures which strike N. 35° E. and dip 80° SE. Along both sets some alteration of the andesite has taken place, the rock being silicified for several inches from the fissure. The ore consists of fragments of the altered rock with some quartz and calcite and carries a little gold, associated with minor quantities of pyrite. The mine is developed by a 200-foot westward crosscut, which intercepts two veins, one of each system. The vein that strikes north has been followed about 300 feet and the fracture that strikes N. 35° E. has been followed for 150 feet. They are quite similar, being in most places an inch or so in width, though in some places they open to 2 or 3 feet. The filling is only partly cemented. A shaft at the mouth of the crosscut is said to be 100 feet deep and to have a westward crosscut at the bottom in which two small veins are opened.

Other prospects.—About 2 miles south of Wellington there are a few prospects on chalcedonic quartz-calcite veins in highly hydrothermally altered andesite. There is a very small amount of copper stain on some of this material, but a large part of the material in the dumps is greenish crushed and altered andesite, the color being due to the presence of chlorite and epidote.

FUTURE OF THE DISTRICTS.

The veins in the quartz monzonite are of small extent and are not strongly mineralized. Some rich pockets of gold ore have been found at the surface, but it would seem that with depth these veins would hardly pay for development. The ore bodies in the andesites have so far not shown any great promise of size. The ore in the Pine Nut Consolidated veins is not very rich and the veins are small. At the Ruby Hill mine the quantity of copper at the surface is probably greater than at depth, and it is problematic if any large ore body will be found.

The most promising ore deposits seen are those in the Triassic(?) sediments at the mouth of Red Canyon and on the ground near the Winters mine.

BATTLE MOUNTAIN DISTRICT, LANDER COUNTY, NEV.

LOCATION AND ACCESSIBILITY.

The Battle Mountain district (No. 7, Pl. I, p. 18), in the northwest corner of Lander County, covers the isolated group of mountains south of Humboldt River and west of Reese River valley. Mining is largely confined to the southeast side of the range, though some prospects not visited during this reconnaissance are located on the northwest flanks of the Battle Mountains. The principal camps of the district visited by the writer, named from north to south, are Copper Basin, Cottonwood Creek, Galena, Copper Canyon, and Bannock. (See Pl. X.)

The mines at the south end of the district are situated in T. 31 N., R. 43 E. of the Mount Diablo principal meridian. The properties at the north end of the district are situated in the southeast part of T. 32 N., R. 43 E., and the southwest part of T. 32 N., R. 44 E.

Battle Mountain, on the Southern Pacific Railroad, is 5 to 8 miles east-northeast of the mines of Copper Basin and Cottonwood Creek, 15 miles northeast of Galena, 18 miles northeast of Copper Canyon, and about the same distance north-northeast of Bannock. Good roads traverse the west side of the Reese River valley southward from Battle Mountain and skirt the south end of the range to Fish Spring valley on the west side of the mountains. Branches from the main road run up the canyons to the mines. Most of these roads are in good repair and hauling is comparatively easy over all of them.

Battle Mountain is the principal supply point for all of northern Lander County and a terminus of the Nevada Central narrow-gage railroad, which extends 98 miles south to Austin. Galena, on Duck Creek, was the largest camp in the vicinity of the mines for many years, but in August, 1912, the camp at the Glasgow Western mine in Copper Canyon, about 3 miles south of Galena, had the largest population. Bannock, at the southeast corner of the mountains, was deserted, and only a few men were living in the vicinity of the Little Giant mine in the northern part of the district.

TOPOGRAPHY.

The Battle Mountains are a rather rugged, isolated group of hills about 16 miles long, from north to south, and 12 miles wide, surrounded by broad, nearly level valleys. Humboldt River flows around the northern end of the range, which apparently deflected the course of the stream, as it makes a sharp bend to the northwest east of the mountains. The Humboldt Valley is exceptionally fertile. East and south of the mountains are the desert, Reese River valley, and a broad westward arm that joins the Fish Creek valley.

Deep canyons have been cut into the mountains from all sides, mostly in radial arrangement, though at the south end of the mountains the drainage lines all run nearly north and south, parallel to the structure of the sedimentary beds in this region.

On the east side of the mountains the heads of the canyons also tend to open out along north-south lines parallel to the structure.

Antler and Sue peaks, in the center of the mountain mass, attain elevations of 8,433 and 8,477 feet, respectively, being about 4,000 feet above the surrounding valleys. The elevation of Battle Mountain is about 4,534 feet, and Galena, on Duck Creek, is about 5,900 feet above sea level.

GEOLOGY.

GENERAL FEATURES.

The geologists of the Fortieth Parallel Survey describe the Battle Mountains as—

for the most part made up of heavy beds of dark quartzites and quartzitic schists, slates, sandstones, and clay beds, overlaid by beds of dark bluish-gray limestones. * * * The volcanic outbursts of the Battle Mountains occupy a comparatively small area. * * * Both rhyolites and basalts break out along the foothills * * * and in the center of the uplift.[1]

---

[1] U. S. Geol. Expl. 40th Par., vol. 2, pp. 667, 671, 1877.

U. S. GEOLOGICAL SURVEY

GEORGE OTIS SMITH, DIRECTOR

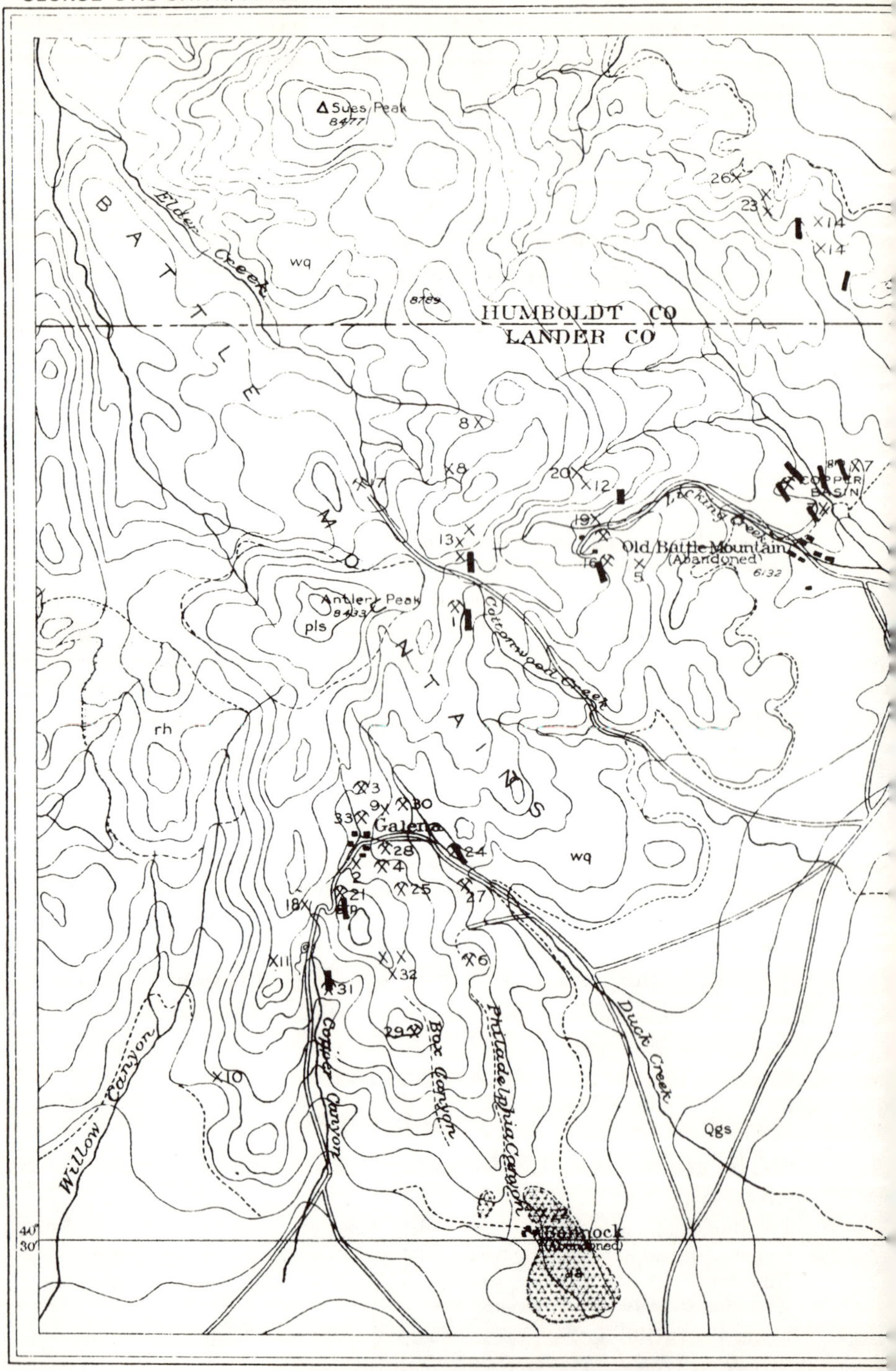

SKETCH MAP OF THE SOUTHEASTERN PART OF T

LANDER AND HUMBOLDT COUNTIE

Topography and geology, with a few changes, from map 5 of the

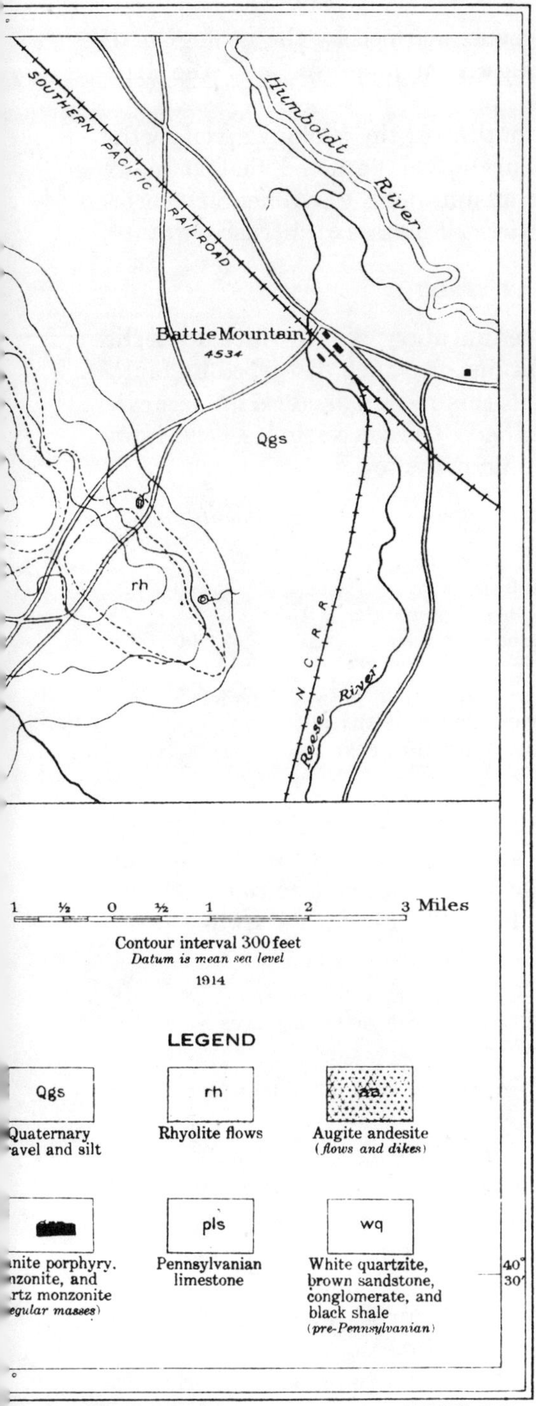

LE MOUNTAINS,
A
el atlas

LIST OF MINES AND PROSPECTS

1 ANTIMONY KING
2 AVALANCH
3 BATTLE MOUNTAIN
4 BRYAN
5 BUCKINGHAM (WESTERN MIN. CO.)
6 BUZZARD (IRON CANYON MINE)
7 COPPER BASIN GROUP (GLASGOW WESTERN MIN. CO.)
8 DEWITT GROUP
9 DRISCOL
10 ELDORADO MINING CO.
11 ELKO-LANDER
12 FAIRVIEW
13 FULLER
14 GOLDEN ERA (SNOW GROUP)
15 COPPER GLANCE GROUP
16 LITTLE GIANT
17 LUCKY STRIKE
18 MEHGER
19 MIDLAND (WESTERN MIN. CO.)
20 MISS NEVADA (WESTERN MIN. CO.)
21 NEVADA MINING CO.
22 NEVADA-OMAHA MIN. CO.
23 NEVADIAN GROUP
24 NORTH BUTTE
25 PLUMAS
26 ROSE SPRING
27 SIOUX
28 SPANISH
29 TOMBOY
30 TRINITY
31 VIRGIN AND SUPERIOR
32 WESTERN LOCK
33 WHITE AND SHILO

The distribution of the formations as mapped by the geologists of the Fortieth Parallel Survey is shown on map No. 5 of the atlas which accompanies their report.

The writer considers the general outline of the geology given by the Fortieth Parallel Survey to be accurate, but he noted that at many places on the southeast side of the mountains the sedimentary series was intruded by small irregular ill-defined masses of intrusive granite porphyry.

SEDIMENTARY SERIES.

Character and thickness.—The sedimentary series which underlie the greater part of the Battle Mountains are much disturbed by faulting and folding, which precluded, in this reconnaissance, an accurate determination of the geologic section. The succession seems, however, to be about as follows, beginning at the top:

Section of sedimentary rocks in the Battle Mountains, Lander County, Nev.

	Feet.
Thin and thick bedded, medium to dark gray, buff weathered limestones (Pennsylvanian)	2,000
Medium and fine grained, red-brown conglomerates with red and yellow micaceous sandstones at top	1,500
Vitreous white quartzite, weathers in red and brown colors	900–1,000
Black, argillaceous, and arenaceous shales, with some interbedded red sandstones near top; thickness unknown but probably great.	

Limestones.—The limestones are well exposed on Antler Peak, and a small area underlain by these rocks occurs west of the rhyolite cap rock at the east side of the mountains west-southwest of Battle Mountain. Of the limestones of Antler Peak the Fortieth Parallel geologists say:

> They extend from the summit to the very bottom of Willow Canyon * * *, exposing 1,200 feet of heavily bedded, dark-gray limestones in places somewhat shaly and of light bluish-gray tints.[1]

Carboniferous fossils were found near the base and 100 feet below the summit of these beds.

The writer collected some fossils from the limestones exposed on the east-west ridge at a point 7 miles west of Battle Mountain, under the rhyolite cap. The three lots collected were examined by G. H. Girty, of the Survey, who reports the following species: *Ambocœlia planiconvexa*, crinoid stems, Productus? sp., Chonetes aff. *C. geinitzianus*, which he says are "probably Pennsylvanian, though the faunas are too limited to be really diagnostic."

Red sandstones.—The red and brown sandstones and conglomerates form a broad northward-lying belt along the central part of

[1] U. S. Geol. Expl. 40th Par., vol. 2, p. 670, 1877.

the eastern slope of the mountains, being the prevailing rocks in the vicinity of Galena and showing in great force in Copper Canyon and the ridge east of that canyon. As a rule the eastward lower beds of this series are arenaceous shales and sandstones of light red and yellow color. The lower central part of the series is conglomeratic. The well-rounded pebbles are largely white quartzite and a dense red jasperoidal material. They range from one-eighth to one-half inch in diameter, though a few larger cobblestones occur. The matrix is a yellowish, red-weathering sand of rather coarse texture. Interbedded with the conglomerates there are numerous irregular, lenslike masses of red sandstone. In the upper part of this series the material becomes fine grained, passing into dirty brown sandstones that contain considerable mica.

White quartzite.—The ridge between Copper Canyon and Willow Creek, at the southwest side of the mountains (see Pl. X, p. 64), is composed of a white vitreous fine-grained quartzite, somewhat similar to the hill north of Cottonwood Canyon in which the Little Giant vein occurs. At the former locality the beds dip west at very steep angles and apparently have been faulted into their present position. At the Cottonwood Canyon exposures the beds dip to the east-northeast at medium angles and are overlain by some conglomerates and light-colored shales which underlie the limestones from which the writer collected fossils.

The two formations discussed above (the red sandstone and the white quartzite) are presumably to be correlated with the Weber quartzite of the Fortieth Parallel Survey reports.

Black shales.—The black shales are exposed in a broad belt along the east front of the range, but this belt narrows abruptly south of Duck Creek, apparently passing under the wash of Reese River valley. At the antimony mines in Cottonwood Canyon (No. 1, Pl. X), a sharp, narrow anticline brings the shales to the surface. In all places they weather a dirty dark brown, but fresh surfaces are black, though in most places the shales are cut by numerous stringers of white calcite. The shales have suffered a great deal of crushing and the normal thin bedding is in most places obscured by a later complicated, somewhat crenulated banding. The shales vary from very fine grained argillites to fine-grained sandstones. They are practically all composed of siliceous material with more or less carbonaceous matter. In places they have a strong petroleum smell when struck with a hammer.

Quaternary sediments.—The youngest sedimentary rocks are the partly consolidated gravels and sands which fill the broad level valleys surrounding the Battle Mountains to an unknown depth. On the east and south sides of the mountains there are few well-marked alluvial cones such as are usually seen in the valleys filled by Quaternary wash. In place of the well-marked cones there are relatively level slopes of low even grade extending from the foothills for long distances into the valleys, the material becoming finer and finer as a greater distance from the mountains is reached. Along the central parts of Reese River and Humboldt valleys there are large areas of fine silts and locally small areas of fine gravels.

IGNEOUS ROCKS.

INTRUSIVE ROCKS.

General features.—There are two overlapping types of intrusive rocks on the east and south sides of the Battle Mountains in the region covered by this reconnaissance.

The intrusives occur both as dikes and sheets. Few of the masses are over 100 feet wide and most of them are 10 to 15 feet across. As a rule they do not extend for more than 200 feet along the strike and are traced with difficulty even in fairly good exposures. Intrusive rocks were seen in all the sedimentary formations except the dense white vitreous quartzite, though with more careful work they would doubtless be found in that formation. The intrusives seem to be more abundant in the lower part of the sandstone conglomerate series than in the underlying dark shales.

The dikes have been cut by some of the veins of the district, and so appear to be older than the veins. There is really no clue to their age, though they probably are related to the late Cretaceous or early Tertiary period of volcanism.

Granite porphyry.—The most widely distributed type found at many places in all the regions visited is a very coarse granite porphyry, containing large white quartz phenocrysts and white or greenish altered feldspars, the largest one-half inch across. The rock might easily be mistaken for one of the conglomeratic beds of the sedimentary series were it not for the fine granular groundmass, which shows distinct feldspar, quartz, and biotite crystals to the unaided eye. The groundmass of these rocks always exceeds the phenocrysts in amount, though the phenocrysts are the most conspicuous characteristic of the type. When examined in thin section these rocks are seen to have two generations of phenocrysts, the large ones being the older. They are quartz and orthoclase as a rule, though in one rock large altered plagioclase phenocrysts were present. The younger phenocrystic minerals are quartz, orthoclase, brown biotite, green hornblende, and a plagioclase feldspar that has an extinction angle between andesine and labradorite. These minerals occur as small grains and partly formed crystals, not much larger than the quartz and orthoclase which constitute the groundmass of the rock.

Alteration of the granite porphyry.—On weathering the groundmass of the granite porphyry disintegrates more rapidly than the quartz and orthoclase phenocrysts, which stand out in high relief, particularly the former mineral. The orthoclase phenocrysts become dull pinkish white in some dikes, apparently due to a kaolinization of the mineral. In other places they turn greenish, the color apparently being due to the formation of some sericite and a yellowish-green chloritic material.

In the vicinity of veins the rock alters to a soft yellowish-white mass of quartz and sericite, the quartz phenocrysts alone remaining to give a clue as to the original nature of the rock. In extreme cases of alteration near veins the granite porphyry becomes completely kaolinized, so that very few of the original quartz phenocrysts remain, the large majority of them apparently having been either crushed or removed by solution.

Monzonite and quartz diorite.—In the vicinity of Copper Basin the second type of intrusive igneous rock is best seen. Though no actual gradation in a single dike was found, different dikes show all gradations from the normal granite porphyry to this rather equigranular quartz diorite porphyry.

The typical monzonite is a medium dark-gray feldspathic rock composed of about equal amounts of orthoclase and plagioclase, together with brownish mica, augite, and some green hornblende. A little quartz is usually seen in thin sections between the feldspar crystals. On the Glasgow western claims in Copper Basin the monzonite passes into a dark-colored quartz diorite, consisting essentially of altered plagioclase feldspars and augite.

The quartz diorite porphyry is a light-gray rock showing to the unaided eye crystals of orthoclase, plagioclase, and biotite. The large percentage of quartz seen in the groundmass would not be suspected from an examination of the hand specimens from the finer grained dikes.

Alteration of the monzonite and diorite.—The monzonite and diorite masses in the vicinity of Copper Basin have all suffered more or less complete alteration, presumably by mineralized solutions which have deposited copper ores in the contact-altered sediments near the intrusive masses. The igneous rocks contain much disseminated pyrite and the thin sections examined all show a large amount of secondary epidote, a very light colored hornblende, and some chlorite. These minerals occur in large irregular areas, apparently having replaced all the original minerals of the rock. The feldspars, where not replaced by the minerals mentioned, are sericitized, and in one section kaolin is extensively developed. Small veinlets of calcite seam the altered igneous rocks at many places.

EXTRUSIVE ROCKS.

Rhyolite.—Two areas of extrusive igneous rocks were seen during this reconnaissance. The ridge trending east and west about 4 miles south of Battle Mountain station is capped by a dark-gray glassy porphyry, which shows distinct flow structure. In thin section the groundmass of this rock is seen to be a rather dark glass in which abundant small crystals of quartz, orthoclase, and a small amount of albite give the light color to the rock. The very dark brown biotite crystals are usually small, but some of them attain a diameter of one-eighth of an inch.

The limestones immediately under this rhyolite are baked to a brownish-red color for 2 to 4 feet below the contact.

The rhyolite west of Willow Creek, shown on Map V of the Fortieth Parallel Atlas, was not visited by the writer. It has been described[1] as "a compact microfelsitic rock and in places develops a tendency to lithoidal texture; in color reddish gray and pink are the prevailing tints, with irregular masses of a deeper purplish red. Individualized minerals, with the exception of quartz, which occurs in large broken translucent fragments, are rare; the feldspars are nearly all fragmentary."

Augite andesite.—The low hills at the southeast corner of the mountains, where the Bannock excitement took place a few years ago, are capped by 50 to 100 feet of greenish-black augite andesite, which is also seen as dikes cutting the quartzite. This rock was called a basalt by the Fortieth Parallel geologists.[2] Bowlders of dark brownish-gray glassy rhyolite, quite similar to the cap rock 4 miles southwest of Battle Mountain, are seen on the surface of the augite andesite area.

STRUCTURE.

The structure of the Battle Mountains is complex. In the short reconnaissance on which this paper is based only the major features were noted.

In the northeast part of the mountains, in the limestone ridge capped with rhyolite (see Pl. X), the beds dip to the east at steep angles near the west end of the rhyolite but have a quite gentle eastward dip at the east end of the ridge. At the base of the limestone occurs a narrow belt of quartzitic conglomerate, which is succeeded farther west by sandstone and fine conglomerate beds that still dip to the east. On a line with the Little Giant and Miss Nevada mines

[1] U. S. Geol. Expl. 40th Par., vol. 2, p. 671, 1877.
[2] Idem, p. 672.

the sediments dip west at medium angles, and near the crest of the mountains at the head of Elder Canyon the siliceous sediments still dip to the west.

At the Antimony King mine in Cottonwood Creek (No. 1, Pl. X, p. 64) the underlying black shales are brought to the surface in what appears to be a close anticlinal fold not over 2,000 feet from limb to limb, in the bottom of the canyon north of the shaft. The relations, however, were not entirely clear, and the shales may have been faulted into their present position. They are cut by a 10 to 20 foot north-south dike of coarse granite porphyry.

In the region of Duck Creek the dark shales occupy a broad belt in the lower part of the canyon. At the eastern end they are folded into a rather broad anticline. The beds near the mouth of the canyon dip eastward, whereas those in the vicinity of the Sioux and North Butte mines (Nos. 26 and 24, Pl. X, p. 64) dip westward at steep angles. Between the mines just mentioned and Galena there is apparently a fault that strikes north-northwest and is entirely in the sandstone conglomerate series. Along this fault the vertical movement has not been of very great magnitude. This fault runs near the Trinity and Plumas mines (Nos. 30 and 25, Pl. X). Near Galena there is a fault that strikes north under the two peaks west of the town. This fault continues southward on the west side of Copper Canyon and to the north apparently swings west of Antler Peak. East of the fault the beds dip to the west at fairly low angles, but west of it the heavy white quartzite beds dip 70°–85° W.

ORE DEPOSITS.

HISTORY AND PRODUCTION.

The Battle Mountain mining district was organized in April, 1866, though it is evident from volume 1 of the district recorder's books that prospectors had been in the mountains for some years previous, as in the minutes of the first meeting of the miners it is stated that the claims located by Tanehill, Sinclair, and Heath before this time (April, 1866) belong to those men. The first claims recorded were the Lake Superior, August 14, 1866, and the Virgin, August 18, 1866. The Avalanche, White, Shiloh, and Humbug were recorded in 1869. The first work in the district seems to have been on the Virgin copper vein in Copper Canyon, though it seems probable that the Little Giant mine was the first steady producer in the district. Hague[1] says that in 1870 the principal mine was the Little Giant, which was first prospected in 1867 and was developed with greater vigor in

[1] Hague, Arnold, Mining industry of central and eastern Nevada: U. S. Geol. Expl. 40th Par., vol. 3, pp. 317–319, 1870.

1868. The production of the mine up to July 1, 1869, according to Hague, amounted to nearly $70,000.

In 1868 the Little Giant and Buena Vista mines were producing, according to the Nevada State mineralogist.[1] By 1870, according to the same authority, 33 mines were in operation[2] and two smelters, having a daily capacity of 12 and 20 tons, were installed at Galena.

In 1872 the antimony deposits in Cottonwood Creek were opened[3] on the Columbia and Mountain King claims, now known as the Antimony King mines.

In 1874 the Battle Mountain district, which had formerly been included in Humboldt County, was ceded to Lander County. The copper mines were worked to some extent by an English company, but little was done with the silver mines, and nothing on the antimony prospects.[4]

In 1876 the English company erected, presumably on Willow Creek, a 30-ton concentrating mill, which was equipped with jigs for separating the copper oxide ores. Prior to this time all the hand-sorted copper ore from the Virgin and Superior mines had been shipped to England by way of San Francisco.[5]

In 1877 Mr. Whitehill reported that water had been reached in most of the properties at a depth of 150 feet and was materially retarding the activity of the camp.

As far as can be learned, the mines of the Battle Mountain district were active until 1885, at which time most of the properties became idle and remained so, except for small leasing operations, until about 1900, when there was a slight revival of activity. The Little Giant has been worked in a small way most of the time.

In 1909 there was a mild excitement at the south end of the district, caused by the reported discovery of rich gold ore at Bannock. Owing to the influx of prospectors following the announcement of this discovery, several of the mines were worked under lease.

In the winter of 1910 some gold prospects were opened in the rhyolite area southwest of Antler Peak (see Pl. X, p. 64), and during the next spring a little work was done on several claims. In 1912 the properties were inactive. It is reported that the ore contains free gold combined with a small amount of silver and some of the gold is associated with small specks of bismuth telluride, tetradymite. From the presence of these tellurides the camp was named Telluride. The ores are said to occur in hydrothermally altered rhyolites, which near the irregular ore bodies are highly silicified.

---

[1] White, A. F., State Mineralogist of Nevada Rept., vol. 2, 1869.

[2] Idem, vol. 3, 1871.

[3] Whitehill, H. R., State Mineralogist of Nevada Biennial Rept., vol. 4, 1873.

[4] Idem, vol. 5, 1875.

[5] Idem, vol. 6, 1877.

In 1912 the Glasgow Western Co. was reopening the Virgin-Superior ground in Copper Canyon and installing a small copper-leaching plant at the mine. The Buzzard vein was being actively exploited by the Iron Canyon Gold Mining Co., and the Nevada vein in the head of Copper Canyon was being reopened. Work on the Little Giant vein at the north end of the district was carried on by four men. Miners were operating in a small way at several places in the district, but in general the mining industry was not very active.

No figures of the early production of the district can be found. It is estimated that the veins near Galena, now held by the French company, have produced between $4,000,000 and $6,000,000 in silver, lead, and gold. The properties held by the Glasgow Western Mining Co., in Copper Canyon and Copper Basin, have produced several hundred thousand dollars' worth of oxidized copper ores, but the exact amount is not known. The Buzzard vein of the Iron Canyon Mining Co. is estimated to have produced about $80,000 in gold, though that figure may be too high. The Little Giant is thought to have produced over a million dollars' worth of silver bullion.

Since 1902 the production of the district has ranged between $1,500 and $65,000 a year, the output being given in the following table:

Production of the Battle Mountain district, Lander County, Nev., from 1902 to 1911, inclusive.

[Compiled from Mineral Resources of the United States.]

Year.	Ore treated.	Gold.	Silver.	Copper.	Lead.	Total value.
	Tons.		*Ounces.*	*Pounds.*	*Pounds.*	
1902	75		2,880			$1,440
1903	625	$5,180	121,095		1,120	56,108
1904	40	41	2,655		14,569	2,105
1905						
1906						
1907	1,452	1,362	8,645	287,300	22,210	65,706
1908	432	4,559	10,922		96,357	14,402
1909	292	12,854	9,238	354	6,930	18,003
1910	11,745	27,592	51,694	4,558	8,804	56,463
1911	12,540	30,479	56,768	3,791	35,029	62,616
	27,201	82,067	263,897	296,003	185,019	276,843

CHARACTER AND DISTRIBUTION.

TYPES OF MINERALIZATION.

The ore deposits of the Battle Mountains occur for the most part in or along fractures, either simple fissures or rather wide zones of fracturing. As a rule they take the form of simple veins or replacement lodes, though the Superior vein of the Glasgow Western Copper Canyon holdings is a single example of a lodelike contact-metamorphic deposit.

There are four well-marked types of mineralization, the most important of which, in point of production, are the silver-lead lodes. The second most important type comprises the copper-bearing deposits, which are in the vicinity of intrusives. There are two distinct areas of copper-bearing deposits, one at Copper Basin, at the north end of the area studied, and the other at the south end of the mountains in Copper Canyon. The third type is represented by the pyritic gold quartz veins, of which a few occur in different parts of the district, though the southwest corner of the mountains contains more gold-bearing veins than any of the rest of the area shown in Plate X. The fourth type, represented in this district by the Antimony King deposits in Cottonwood Canyon, comprises veins and replacement lodes carrying stibnite. This type is more or less associated with the first type, as in most of the lead-silver deposits some antimony is present, either associated with galena or gray copper or in the form of stibnite.

SILVER-LEAD DEPOSITS.

Distribution.—The silver-lead deposits of the Battle Mountains are found in all the sedimentary rocks but do not appear to bear a very close relation to either the granite porphyry or monzonite intrusives. In the dense quartzite and sandstones the ores, as a rule, occur in simple fissures, but in the shaly beds the ores are found chiefly in zones of fractured and mineralized country rock, where, though they still retain somewhat the form of veins, they have more of the nature of replacement lodes.

Minerals.—The minerals most commonly seen in these deposits are galena, dark (nearly black) sphalerite, pyrite, and tetrahedrite. The proportion of these minerals to one another differs in the different veins, but it seems to be fairly well established that the richer ores carry more tetrahedrite than the less valuable ores, and that when much pyrite is present the silver values are low.

In the oxidized ores near the surface cerusite is the most abundant ore mineral, though some anglesite is usually present, and it is reported that the richest surface ores of the Little Giant, White and Shiloh, and Trinity veins all carried silver chloride, though none of this mineral was definitely determined by the writer. The oxidized zone in most of the silver-lead veins is shallow, in few places reaching a depth of 150 feet and usually being considerably above the 100-foot level. At many places, particularly in the canyon bottoms, the sulphides are found at the surface.

Enrichment.—The zone of sulphide enrichment is apparently very limited, though few data were obtained on this important subject in the district, as so many of the most important veins were not accessible for study. In the Little Giant vein the secondary sulphides polybasite, pyrargyrite, argentite, and tetrahedrite were found in the No. 5 tunnel about 150 feet below the croppings, whereas in the No. 8 tunnel, 150 feet lower, the ores were the originally deposited sulphides. A little argentite was also found in cracks in primary galena and sphalerite from ores about 75 feet below the croppings of the vein, and it is reported that ruby silver was found in the surface ores.

In this type of deposit the valuable ores are localized in shoots of different lengths, though from all accounts none of the shoots worked in the past were very long.

The carbonate silver-lead ores of the district are said to carry between 100 and 200 ounces of silver to the ton on the average, but the primary sulphide ores are usually worth about $40 a ton, including both silver and lead.

COPPER DEPOSITS.

Distribution.—The copper deposits of the Battle Mountains occur in rather small areas. The northern area lies at the west end of the rhyolite-capped ridge southwest of the town of Battle Mountain, at the northeast corner of the mountain group, and comprises Copper Basin and the hills north of the basin, where the Golden Era and Copper Glance claims are located. The southern deposits are all situated near the center and north end of the ridge between Copper and Box canyons.

Copper Basin.—The deposits near Copper Basin are inclosed in light-colored, highly indurated sandstones and siliceous slates that are intruded by irregular dikelike masses of granite porphyry and quartz diorite. The larger deposits lie near a mass of the dioritic rock. The development has been done largely in oxidized ores, malachite, chrysocolla, and azurite being the most common minerals, though cuprite is present. These minerals are most abundant along joint cracks in the rock but have also replaced the minerals of the rocks, particularly the matrix of the coarser beds of sandstones. At one place a shaft was sunk to an unknown depth, probably about 150 feet, in which cupriferous pyrite was found.

Copper Canyon.—The large majority of the deposits of copper ores on top of the ridge east of Copper Canyon are similar to those described near Copper Basin. The Virgin and Superior workings in Copper Canyon are on the most extensive copper deposit in the district. The ores at this place occur along a series of subparallel fractures that strike northward and dip westward in sandstones and siliceous and calcareous shales. West of the ore body lies a dike of granite porphyry. Some of the calcareous beds show the effects of contact metamorphism, being altered to masses of dark-brown garnet and epidote. The siliceous beds, however, show little alteration, though they are more or less mineralized. The copper ores occur through this crushed zone about 150 feet in width and of unknown length north and south. They are entirely oxidized to a depth of more than 113 feet, though at the 188-foot level chalcocite and chalcopyrite ores were encountered in the Virgin ore body. At the same level in the Superior workings the ores consist of chalcopyrite, pyrite, dark sphalerite, and some galena, associated with contact-metamorphic minerals.

These copper ores are slightly auriferous and on the surface some limonitic material in fractures between the Virgin and Superior workings have yielded free gold.

GOLD DEPOSITS.

The principal mine on a gold vein in the Battle Mountain district is the Buzzard. The workings are located at the east end of the summit of the ridge south of Duck Creek and all the ores are oxidized. The vein is narrow, in many places pinching to a tight fracture. The ore, an iron-stained quartz carrying free gold, occurs in short shoots. The Plumas and Bryan mines, located nearer the level of Duck Creek about a mile southeast of Galena, were also producers of gold ore. These mines could not be entered far enough to give an adequate understanding of the veins, but the ore on the dump was mainly white quartz carrying pyrite.

ANTIMONY DEPOSITS.

The Antimony King mines (No. 1, Pl. X, p. 64) are the only representatives of this type of deposit in the district. The mines are located on a steep narrow ridge between two forks of Cottonwood Canyon. The deposits were known in 1871 and have been worked by different people since that time. In 1912 three claims on the strike of the deposit were held by J. M. Pine, of Denver, Colo. The vein occurs along the western side of a tight anticline composed of nearly black siliceous shales overlain by light-colored quartzite. On the eastern limb of the fold there is a 10 to 15 foot dike of coarse granite porphyry, which dips 60° E., cutting the folded sediments at an acute angle. The vein cuts both the quartzite and underlying shales, striking N. 17° E. and dipping 75° W. In the quartzite at the summit of the ridge it is quite regularly 2 feet in width, but where it passes into the shale it broadens to about 8 or 10 feet of mineralized crushed country rock in which there are irregular stringers and pockets of antimony ore. In the quartzite the vein is largely filled with white and dark quartz associated with stibnite, but in the shales the greater part of the vein is crushed silicified shale with stringers and pockets of stibnite and a little pyrite replacing the shattered country rock.

In some specimens from a lower tunnel in the north gulch a small amount of brownish antimony oxide has developed in drusy cavities in the stibnite quartz ore. The property is developed by several shafts on the crest of the ridge, two of which must be between 50 and 100 feet deep, and by numerous open cuts, shallow shafts, and tunnels in the gulches on either side.

PROPERTIES AT NORTH END OF BATTLE MOUNTAIN DISTRICT, INCLUDING COTTONWOOD CANYON, LICKING CANYON, AND COPPER BASIN SUBDISTRICTS.

NEVADIAN GROUP.

The Nevadian group of claims (No. 23, Pl. X, p. 64) lies at the north end of the district, on the northeast side of the Battle Mountains. Some development has been done on a number of fractures that strike N. 10°–15° W. in light-colored sandy shales and quartzites, which dip west at fairly low angles. The principal work is a 75-foot incline shaft on a vein that ranges from a fraction of an inch to 10 inches in width and is filled with rusty crushed rock and a little quartz showing casts of pyrite. The ore is said to carry from $1 to $3 a ton in gold.

ROSE SPRING GROUP.

A short distance northwest of the Nevadian there is a tunnel on the Rose Spring Co.'s ground (No. 26, Pl. X). This tunnel runs S. 10° E. for 400 feet through westward-dipping arenaceous sediments, following a tight, nearly barren fracture that opens out near the face to 4 inches in width and shows a small amount of pyrite.

GOLDEN ERA GROUP.

The Golden Era group of 17 claims (No. 14, Pl. X) is located in low hills about 6 miles west-northwest of Battle Mountain. Two veins of this group have been developed to some extent and numerous prospect pits are located on nearly parallel stringers at many places. A vein near the north center of the group is developed by a 100-foot incline shaft, with short drifts at 50 and 90 feet, and by a 300-foot drift, which cuts the shaft 20 feet below the collar. The vein strikes N. 5° E. and dips 70° E., cutting sandstones and shales that dip west-northwest at low angles. In most places the fracture is 2 to 4 feet in width and is filled with iron-stained crushed country rock that in places shows small bodies of white quartz. In the shales the simple fracture splits into a number of subparallel slips. At the bottom of the incline the fracture penetrates a bed of dark shales in which there is some pyrite, the only sulphide seen in this vein. The small quantity of gold is reported to be entirely free.

A vein about one-fourth of a mile south of that last described is opened by a 150-foot drift and a shaft of unknown depth, now full of water, which is equipped with a 15-horsepower Fairbanks-Morse gasoline hoist. The vein strikes N. 10° E. and dips 75° W., cutting highly indurated sandstones, some of the beds being very coarse grained. It ranges from 2 inches to 2 feet in width in the drift, the filling being largely crushed iron-stained sandstone, with some stringers and lenses of quartz containing a little pyrite.

COPPER GLANCE GROUP.

The Copper Glance group of 22 claims (No. 15, Pl. X) is located at the western end of the east-west rhyolite-capped ridge, at the northeast side of the Battle Mountains, about 4 miles west of Battle Mountain. The chief development, a 150-foot vertical shaft and drift, has been done on a body of oxidized copper carbonate ores at the north end of the group. The ores occur along a fracture zone that strikes N. 80° W. and dips northward in light-colored dense arenaceous shales and interbedded sandstones that strike N. 45° E. and dip 60° SE. The principal ore body lies at the surface, about 60 feet south-southeast of the shaft, where a mass of broken shales, about 20 feet wide and 150 feet long, contains some chrysocolla, copper carbonate, and copper pitch ores between the fragments of country rock. A crosscut at a depth of 100 feet in the shaft runs south for 50 feet but shows no ore body, though near the shaft a zone of fractured shales, dipping north at a low angle, shows a very small copper stain.

South of this group, near the sandstone-limestone contact, there are a few copper prospects in arenaceous beds adjacent to a dike of granite porphyry.

GLASGOW WESTERN COPPER BASIN CLAIMS.

The Copper Basin claims of the Glasgow Western Mining Co. are located north of the old abandoned town of Battle Mountain, shown in the north center of Plate X (p. 64). There are 47 claims in the group, on most of which there are some copper prospects. This basin is underlain by quartzitic shales and some beds of limy sandstone that have been folded to a slight extent. On the east side of the group the beds dip east-northeast at medium angles underlying the Pennsylvanian limestone under the rhyolite hills. On the western side the beds dip to the west at low angles. The crest of the anticline runs north-northwest. Along the crest the sediments have been intruded by a gray feldspathic porphyritic quartz diorite that possibly is a basic phase of the granite porphyry and monzonite intrusion of the region. The igneous rock has been altered by the addition of a large quantity of epidote and a very light greenish amphibole. The epidote has replaced the femic minerals of the diorite, but the amphibole appears in large areas of the slides, apparently replacing all the original constituents.

Along the western flank of the fold there are dikes and sheets of granite porphyry, similar to the usual type seen at many localities in the Battle Mountains. Along both types of igneous rock, but particularly near the quartz diorite, there are narrow zones of contact metamorphism ranging from 1 foot or 2 feet to a maximum of 20 feet in width. In the contact zone there is more or less copper in the form of the carbonates and silicate, associated with iron oxide and some manganese oxide. At several pits beds of coarse gritty sandstone, apparently near no intrusive bodies, are copper bearing. The carbonates, mostly malachite, replace the calcareous matrix of these beds. In the fine-grained sandstones and quartzitic beds the copper minerals occur in joints and locally replace the rock for short distances from them. In some of the ore there is a small quantity of red oxide, cuprite, which occurs as thin films along joint cracks and is usually partly altered to the carbonate minerals.

At a few places where development has reached a greater depth than 50 feet pyrite, which may be slightly cupriferous, occurs in fine grains disseminated through the sandstone and as small veinlets along what appear to be joint planes.

The development work on these claims consists of shallow open cuts and pits for the most part, though the ground has been prospected by diamond drills and a deep shaft is located near the south end of the basin about one-fourth of a mile east-northeast of the old town. The Blackbird tunnel, in the west center of the group, is a crosscut which extends 700 feet N. 60° W. and attains a maximum depth of 150 feet. It runs through barren ground, largely quartzite cut by some dikes of granite porphyry.

WESTERN NEVADA GROUP.

The Western Nevada Mining Co. owns 18 claims in the lower part of Licking Canyon, about 2½ to 3 miles west of old Battle Mountain. The group is 3,600 feet wide from east to west and 4,200 feet long, the Midland vein being located near the center of the group.

Miss Nevada vein.—The Miss Nevada crosscut (No. 20, Pl. X, p. 64) is on the south side of the north fork of Licking Canyon. It runs S. 50° E. for 130 feet through quartzites that dip west at medium angles. Four or five small quartz pyrite veins are disclosed by the tunnel, and one veinlet carries molybdenite in the joints of the wall rock for a few inches on either side of the vein. Near the face of the tunnel there is a 25-foot winze on a vein that strikes N. 30° E. and dips 40° WNW. This fracture, about 2 feet in width, is filled with crushed, iron-stained country rock, with a few stringers of white quartz and cupriferous pyrite. Water stands in the winze 10 feet below the collar, and some of the pyritic ore at the water level is coated with thin films of chalcocite. This ore is said to carry free gold and some silver.

At several places between the Miss Nevada and the Midland veins (Nos. 20 and 19, Pl. X) there are shallow prospects on veins that strike N. 30° E. and dip westward, some of which carry principally gold and others largely silver-lead. About 800 feet north of the Midland there is a brownish and yellowish stained mass of altered granite porphyry, which is said to carry silver values, though as far as developments show pyrite is the only sulphide in the rock.

Midland vein.—The Midland vein (No. 19, Pl. X) of the Western Mining Co.'s holdings is opened by a 250-foot drift tunnel and a 98-foot shaft, now under water, located on the south side of the south fork of Licking Canyon, about 3 miles above the old town of Battle Mountain. The vein strikes N. 10° W. and dips 80° W., cutting a light brownish-gray quartzite which is heavily impregnated with very fine grained pyrite in the vicinity of the vein. It ranges from a tightly frozen fracture to 18 inches in width. The wider portions, filled with quartz and sulphides, constitute the ore shoots that stand vertical or pitch to the north at steep angles. The shoots are short, few of them being over 20 feet in length and most of them between 10 and 15 feet. Three shoots are exposed in the tunnel. The ore minerals are galena, dark sphalerite, arsenopyrite, pyrite, tetrahedrite, and chalcopyrite. A small quantity of a soft black mineral that seems to be chalcocite occurs in fractures in the ore from the tunnel. None of the rich silver-antimony minerals were noted, but they may be present. The sulphides occur immediately at the surface at the mouth of the tunnel. The ore is said to carry about $18 in silver, $1.50 in gold, $5 in copper, $10 in lead, and $16 in zinc to the ton. The quantity of silver differs according to the copper content of the ore, though the galena is also argentiferous.

Several shipments of lead-silver ore have been made from this vein, which is the most extensively developed on the Western Mining Co.'s ground.

Northland vein.—The Northland vein seems to be the northward continuation of the Midland vein, though in the 200-foot drift and in the 100 feet of the 150-foot incline which could be entered the vein strikes N. 10° E. and dips 30° W. The ore is oxidized to a depth of 25 feet, below which the vein is filled with crushed wall rock, with abundant disseminated pyrite and stringers of white quartz containing cupriferous pyrite and arsenopyrite. No lead or zinc sulphides were seen.

Fairview vein.—The Fairview vein (No. 12, Pl. X) of the Western Mining Co.'s group lies on the north side of the south fork of Licking Creek. The vein strikes N. 11° W. and dips on the average 50° W. It is a fracture zone which ranges from 3 to 16 feet in width, being almost entirely filled with crushed quartzite, the country rock. At a depth of 75 feet in the 225-foot incline shaft the first sulphides show in the ore. Pyrite, which seems to be slightly cupriferous, as it is in some places coated with thin films of chalcocite, occurs in stringers 4 to 8 inches wide near the hanging wall of the fracture and is more or less widely disseminated throughout the remainder of the 8 feet of the crushed zone. The sulphide is said to run about $12 a ton in gold, silver, and lead, though no galena was noted in the ore examined. Above the sulphide zone the fracture filling is stained with iron oxide and a yellowish powdery substance that contains both iron and lead, and may possibly be a mixture of the sulphates of those metals.

Buckingham deposit.—The Buckingham deposit at the south end of the Western Mining Co.'s holdings in Licking Creek (No. 5, Pl. X) is a northerly striking replacement of a bed of dark quartzite which dips 38°–40° W. The hanging wall is a well-marked fracture filled with gouge, below which for 12 to 15 feet the quartzite contains more or less disseminated pyrite cut by small stringers of tetrahedrite, sphalerite, and galena for 3 feet below the hanging wall. The hanging-wall ore carries considerable silver, that at the bottom of the 95-foot incline averaging $14.75 a ton in gold and silver, the latter metal predominating. This deposit is developed by the incline and by a crosscut 140 feet in length, which cuts the incline 69 feet below the collar. At the tunnel level there are 130 feet of drifts in the ore zone.

LITTLE GIANT MINE.

The Little Giant mine (No. 16, Pl. X, p. 64) lies on a vein which strikes N. 30°–60° W. and dips southwest at medium angles. The vein was opened in 1867 and has been worked rather continuously since that time. Hague[1] reported that in 1869 the vein had been stripped across the top of the hill and was developed by three tunnels in the gulch to the west. He reports that the oxidized surface ores carried from $100 to $1,000 a ton in silver and were easily treated by amalgamation, but that the primary ore contained lead, antimony, and other base metals from which it was difficult to extract the silver in the 5-stamp mill that was located near the mouth of the canyon.

The hill in which the Little Giant is located is composed of quartzite and light-colored siliceous shales. On the west side of the hill

[1] Hague, Arnold, Mining industry of central and eastern Nevada: U. S. Geol. Expl. 40th Par., vol. 3, p. 318, 1870.

a mass of intrusive granite porphyry is exposed at the mouth of the lower west tunnel No. 8 and along the road to the mill.

In 1912 there were eight tunnels, two shafts, and two long, open cuts or underhand stopes from the surface along the Little Giant vein for about 1,700 feet of its length. The longest five tunnels are situated in the gulch west of the summit of the hill. The largest tunnel, No. 8, shows about 500 feet of the vein and reaches a depth of about 250 feet below the croppings. At a depth of 200 feet the ore, though slightly oxidized, consists largely of galena, dark sphalerite, pyrite, and a little tetrahedrite.

The vein in this tunnel cuts both the quartzite and the porphyry, being narrower in the porphyry. In the sedimentary rocks the vein commonly expands to a width of 8 feet, consisting of crushed rock with many narrow bands of sulphides throughout its width.

Most of the tunnels lie in the oxidized portion of the vein, where the siliceous ores are coated with a yellowish powder, consisting largely of hydrous iron oxide but carrying lead and silver, probably as cerusite and silver chloride. In this ore there is some galena and anglesite. In the seeming completely oxidized ore there are in many places remnants of sulphides, chiefly galena, surrounded by anglesite, but argentite and polybasite are both present in lesser quantity.

In one of the central tunnels, No. 5, the main stopes are situated in secondary sulphide ores. At this place the vein, which lies in light-colored siliceous shales and quartzites, has a steplike arrangement. The flat portions dip 15°–20° SW., and in them the ore is 8 to 10 inches wide and is very rich, consisting of porous quartz with secondary tetrahedrite, polybasite, pyrargyrite, argentite, and pyrite. Where the vein pitches 50° SW., cutting the bedding of the wall rock, the ore makes through a zone 6 to 8 feet wide and consists of country rock with abundant disseminated pyrite cut by numerous interlacing veinlets of pyrite and by later veins carrying secondary silver sulphides. The rich ore from the "flats" is said to average about 150 ounces of silver to the ton, but the ore from the steep pitches is much lower in grade.

The Little Giant mill is about one-half mile north of the mouth of tunnel No. 8, near the south fork of Licking Gulch. It is equipped with five stamps and Diester sand and slime tables. It is said, however, that the recovery from this mill is very unsatisfactory, owing to the excessive sliming of the high-grade ores.

FULLER GROUP.

The Fuller group (No. 13, Pl. X, p. 64) is on the north side of the north fork of Cottonwood Canyon, north of the Antimony King property. The fold which has exposed the dark shales at the King mine dies out north of the canyon in a low swell, so that the country rocks of the Fuller group are the sandstone and shale of the overlying series. They are cut by a northerly striking granite porphyry dike, similar to the dike at the Antimony King.

Several narrow iron-stained quartz veins are exposed in the surface workings of the group, which are said to carry gold usually associated with a small amount of silver and ranging in value from $7 to $150 a ton. At the one tunnel which reaches the sulphides, about 25 feet below the surface, pyrite is the only metallic mineral. In another vein a little residual galena was noted, but as a rule the ores are typical pyrite gold ores.

DEWITT CLAIMS.

The Dewitt claims (No. 8, Pl. X, p. 64) are on the summit of the range at the head of Elder Creek, about 1¾ miles north of Antler Peak. The country rock is a light-colored, highly indurated sandstone, nearly quartzite, in which there are a series of small quartz veins that strike N. 10°–15° E. and dip eastward. There are shallow pits on a number of the veins and a short tunnel in what appears to be one of the promising locations. The sandstone near the veins is somewhat altered, chiefly by a sericitization of the matrix. A little pyrite and arsenopyrite are disseminated in this rock. The vein filling is iron-stained quartz and crushed country rock. The development is as yet entirely in oxidized ores, some of which show lead carbonate stains. The ore is said to contain gold and silver, but the grade is not known.

LUCKY STRIKE MINE.

The Lucky Strike mine (No. 17, Pl. X) is on the west side of Elder Canyon near its head, about 1½ miles north of Antler Peak. Three of the five claims are located on the vein, which strikes N. 35° W. and stands nearly vertical or has a steep westward dip in quartzite. The development consists of a 70-foot crosscut to the vein, which is cut at a depth of 60 feet, and 400 feet of drifting at the tunnel level; there is also a 106-foot shaft near the point where the vein is cut by the tunnel. All the work is inaccessible. At the surface the vein is 2½ feet wide and consists of nearly solid galena and dark-colored sphalerite, showing only slight oxidation. The ore taken from the tunnel level is entirely unoxidized and is said to average about 50 per cent of lead, 150 ounces of silver, and $4 to $5 in gold a ton. Several small shipments were made from this property about 1908.

PROPERTIES AT SOUTH END OF BATTLE MOUNTAIN DISTRICT, INCLUDING GALENA, COPPER CANYON, AND PHILADELPHIA CANYON SUBDISTRICTS.

SIOUX CLAIMS.

The Sioux claims, three in number (No. 27, Pl. X, p. 64), are on lower Duck Creek, about 1½ miles east of Galena. The development work consists of two tunnels on the vein and some open cut and shaft work higher up on the hill south of the valley. The vein occupies a fracture zone in nearly black siliceous shales which strikes N. 15° W. and dips on the average 75° W. In the lower tunnel, 200 feet long, the vein is about 4 feet wide, and in the 120-foot upper tunnel it ranges from 4 to 6 feet in width. The filling is largely crushed black shale containing irregular stringers and bunches of white quartz carrying pyrite. The crushed rock carries irregular masses of sulphides, largely dark sphalerite and galena with some pyrite and possibly tetrahedrite. The value of the ore is not known, as no assays or shipments had been made from the property previous to August, 1912.

NORTH BUTTE VEIN.

The North Butte vein (No. 24, Pl. X, p. 64) is developed to a depth of 75 feet by a crosscut on the north side of Duck Creek about 1 mile below Galena. The ore body is about 15 feet wide, strikes N. 15° W., and dips 50° W., parallel to the structure of the dark-colored siliceous shales. Immediately below the ore there is a 15 to 20 foot dike of granite porphyry, which has been very much altered and in some places mineralized.

Throughout the ore zone the crushed country rock contains abundant disseminated pyrite. That mineral, together with galena, dark sphalerite, and a little cupriferous pyrite, is seen in veins associated with white quartz that cut the ore zone on either the footwall or hanging-wall side of the vein. In one specimen of ore argentite was found in cracks in the other minerals, and it is reported that some ruby silver was discovered in ores which were mined from surface cuts.

BUZZARD VEIN.

The Buzzard vein (No. 6, Pl. X, p. 64) is developed by open cuts, a shaft, and drift tunnels on the top of the ridge south of Duck Creek, about 2 miles southeast of Galena. This property is one of 19 claims belonging to the Iron Canyon Gold Mining Co., and shows the most extensive development work. The country rock is light-colored, fine-grained quartzite, the beds dipping 20° W. The vein strikes N. 10° W. and dips 65°–70° W. It ranges from a tight fracture to 4 feet in maximum width. The ore occurs in short shoots, 30 to 75 feet in maximum length, that are parallel to the dip of the vein and in its widest portions. The ore is porous yellowish limonitic-stained white quartz, though a little pyrite and a very few chalcopyrite fragments are seen in ore from the tunnel levels about 150 feet below the outcrop. The surface ore is said to have carried as high as $120 in gold a ton, but at the tunnel level the average value is $17 a ton.

The ore is skidded down a very steep trail north of the tunnel to a 20-ton cyanide mill about three-fourths of a mile northwest of Wilson ranch at the mouth of Duck Creek. The ore is crushed in a 4-foot Huntington mill, classified, and the sand and slimes treated in separate leaching tanks.

NEVADA-OMAHA.

The Nevada-Omaha properties at Bannock (No. 22, Pl. X, p. 64) were not being worked in August, 1912. They are located at the extreme southeastern corner of the Battle Mountains, on the east side of Philadelphia Canyon. The dense white quartzite in this vicinity is intruded by dikes of augite andesite, which rock also forms a capping over a small area at and southeast of the properties. A dark brownish-black glassy rhyolite flow overlies the older rocks. The main development is a 350-foot vertical shaft and a 300-foot crosscut tunnel, which intersects the shaft about 75 feet below the collar. The shaft could not be entered, but in the crosscut the augite andesite showed no sign of mineralization aside from thin films of pyrite on some joint planes. It is said that there was a small gash vein in the shaft to a depth of 27 feet, which was filled with white quartz carrying free gold.

FRENCH COMPANY'S HOLDINGS.

The French company's holdings at the town of Galena comprise 22 claims and 3 mill sites. These claims were acquired in the early eighties, and for about five years considerable mining was done on the company's ground. Since 1885 the mines belonging to this company have been idle. It is estimated that the production from the veins now held by the French company is between four and six million dollars in silver and gold.

White and Shiloh vein.—The White and Shiloh vein (No. 33, Pl. X, p. 64) of the French company is on the east side of the north head of Duck Creek, and is opened by two tunnels and a 160-foot shaft, all abandoned and in a caved condition. Water stands very near the collar of the shaft. This vein, which was a large producer in the early eighties, has not been worked to any extent since 1885.

It is reported that the ore has been taken out for 1,300 feet along the vein north of Galena to a depth of 160 feet.

The vein strikes approximately north and south and dips west at medium angles along a broad zone of fracture in light-colored siliceous shales, sandstones, and conglomerates. The ore is said to have occurred as a replacement of the crushed wall rocks where the main fracture splits. It consisted of argentiferous galena and some tetrahedrite with pyrite, and the ores in the main shoots contained rich silver sulphantimonide mineral. The material on the dumps is largely pyrite and some galena, which is said to have come from a depth of about 160 feet, where the ores became too lean to work.

Battle Mountain vein.—The Battle Mountain vein (No. 3, Pl. X, p. 64) of the French company's holdings lies immediately east of the White Shiloh vein on the hill north of Galena. The vein strikes N. 45° W. and stands vertical. It is a zone of fractured fine conglomerate and red sandstone, ranging from 4 to 8 feet wide, and the ore occurs along the northeast wall in a band averaging 2 feet wide. It is shown in the Lady Blain crosscut tunnel that runs N. 70° E. and is caved 250 feet from the mouth. The vein crosses the tunnel about 90 feet from the portal at a depth of about 25 feet. The ore is a reddish-brown soft material carrying some stibnite and remnants of galena surrounded by some anglesite. Cerusite and possibly horn silver are present, and what appears to be argentite replaces some of the galena. It is high-grade silver ore as far as the development shows. Between the Lady Blain and Trinity mines (Nos. 3 and 30, Pl. X, p. 64) there are several veins that strike N. 10° E. and dip 45°–50° W., nearly parallel to the sandstone conglomerate bedding, all showing some oxidized ore at various shallow shafts and pits.

Avalanche claim.—The Avalanche claim of the French company (No. 2, Pl. X, p. 64) is on the east side of the south fork at the head of Duck Creek, 1,000 feet south of town, on the White and Shiloh vein. The development consists of an 800-foot drift tunnel and a raise to a level 80 feet above the drift that commences 600 feet from the portal. The vein is represented in most of the tunnel by a barren fracture, which strikes N. 5°–10° W., dips 60° W., and ranges from 2 to 8 feet wide, being filled with crushed calcareous sandstone and quartzite. At 500 feet from the portal the vein splits. The hanging-wall portion runs into quartzite, where the fracture is strongly marked, but barren as far as developed. The footwall portion continues in the limy sandstone but fingers out near the face of the 800-foot drift. The ore occurred in a 100-foot shoot, 60 feet above the tunnel level, and was 22 feet wide. It consisted of a mass of crushed, very calcareous sandstone, in which there were large irregular pockets and stringers of "sand carbonate," with some remnants of large galena crystals. It is said that the silver content of the ore from this stope was rather low. The shoot continued on the dip of the vein to a point within about 15 feet of the tunnel level and nearly to the surface, about 70 feet above the sublevel.

Trinity vein.—The Trinity vein (No. 30, Pl. X, p. 64) of the French company is about half a mile northeast of Galena. It is developed by an 800-foot drift tunnel that attains a maximum depth of about 250 feet. The vein strikes N. 30° E. and dips 45°–50° W., in black siliceous shales near the contact with the overlying sandstone series. The shales for a width of 15 feet are crushed and contorted and contain abundant disseminated pyrite and arsenopyrite. The ore occurs in irregular lenses and stringers, 1 to 4 feet in width, near the central portion or hanging wall of the crushed ore. It consists of an intergrowth of white quartz, arsenopyrite, pyrite, antimonial silver-bearing galena, and an abundance of black sphalerite, and is said to have averaged about $20 per ton in silver and gold.

DRISCOL PROPERTY.

The Driscol property (No. 9, Pl. X, p. 64), consisting of three claims, lies between the White and Shiloh and the Trinity veins on the hills north of Duck Creek. This property was extensively worked in the late seventies by four drift tunnels through a vertical range of 200 feet. The vein strikes N. 25° E. and dips on the average 50° W. in highly indurated sandstones and fine conglomerates. It ranges from 3 inches to 4 feet in width and is largely filled with crushed country rock. An ore streak between 3 and 4 inches in width follows the hanging wall. It consists of massive galena, some gray copper, and a little sphalerite and pyrite. In all the tunnels which could be entered the ore was somewhat oxidized to a mass of iron-stained cellular cerusite surrounding unaltered galena. Anglesite is present in some of the ore. The galena carries some antimony and silver and the oxidized products are said to be rich silver ores.

SPANISH VEIN.

The Spanish vein is on the south side of Duck Creek about one-fourth of a mile southeast of Galena (No. 28, Pl. X, p. 64). It is from 2 to 10 inches in width, strikes N. 10° E., and dips 75° W., about parallel to the bedding of the sandstone and shale country rock. The ore minerals are galena, sphalerite, and pyrite, the accompanying oxidized ores forming a shallow zone near the surface.

BRYAN VEIN.

The Bryan vein (No. 4, Pl. X, p. 64) is developed by several tunnels and a shaft, all of which were inaccessible except for short distances. These workings are located on both sides of a south branch of Duck Creek about three-fourths of a mile southeast of Galena. It strikes nearly north and dips to the west at steep angles, cutting sandstones which dip 15° W. The fracture ranges from a few inches to 10 feet in width. In the narrower portion it is filled with white quartz carrying pyrite and in the wider portion the filling is pyritized crushed wall rock. The ore, said to have averaged about $12 in gold to the ton, was milled in a 10-stamp amalgamating plant on Duck Creek, about one-half mile northeast of the shaft. The mill had been removed prior to 1912.

PLUMAS VEIN.

The Plumas vein (No. 25, Pl. X, p. 64) is in a side gulch about three-fourths of a mile southeast of Galena and one-half mile south of Duck Creek. It was developed by a shaft and tunnel in the early days of the camp, but the workings are accessible for only a few feet from the mouth of the tunnel. The vein where seen is from 4 to 6 feet wide and lies in sandstones immediately below a bed of dark shale. There has been considerable crushing along this zone, which strikes N. 20° W. and dips 55° W., parallel to the bedding of the sediments. The vein filling is largely iron-stained crushed country rock with large lenses of porous iron-stained quartz, which constitutes the ore. A little residual pyrite is visible in some of the quartz, and more is present in ores from the dump that presumably came from the shaft. The oxidized ore is said to carry from $10 to $12 in gold to the ton.

WESTERN LOCK.

On the Western Lock ground (No. 32, Pl. X, p. 64) there are a number of open pits and short tunnels on a series of short lenslike bodies of copper oxide ores. The ores make along a series of vertical fissures that strike N. 35° W. in highly indurated sandstones and fine conglomerates. Chalcopyrite has apparently replaced the sandstones to a limited extent. The oxidized copper minerals, including chrysocolla, malachite, cuprite, and copper pitch ore, fill joints in the sandstones and in some places replace the sandstone around kernels of altering chalcopyrite.

TOMBOY PROPERTY.

The Tomboy property (No. 29, Pl. X, p. 64) is on the east side of the summit west of Copper Canyon near the head of what is known as Box Canyon. The country rock, coarse and fine sandstone, highly indurated in most places, strikes north and south and dips west at fairly low angles. These beds are cut by a series of fissures that strike N. 20° W. and dip eastward. The gold ore consists of a fine conglomerate, the sandy matrix of which has been altered and replaced to some extent by pyrite in the vicinity of the fissures. In the surface development work the whole mass is stained red and yellow and only small remnants of unaltered pyrite are seen. This bed is about 50 feet thick and is said to carry $7 a ton in gold for a length of about 700 feet along its strike.

NEVADA CLAIMS.

The Nevada mining properties consist of six claims at the head of Copper Canyon near the divide into the south head of Duck Creek (No. 21, Pl. X, p. 64). The main development is a 250-foot vertical shaft filled with water to within 190 feet of the collar. The levels at 60 and 150 feet are very irregular, following open water-courses in steeply westward dipping limestones in which there are pockets of ore. At 300 feet west of the shaft there is a 20-foot dike of monzonite porphyry, which strikes N. 10° E. and dips 45° W., cutting the sedimentary beds at an acute angle. There is little contact metamorphism along this dike, but considerable crushed and iron-stained limestone occurs along both walls. The ore zone at the main shaft strikes N. 15° W. and dips 50° W. Underground it is about 40 feet wide and has been developed for about 150 feet along the strike on the two levels. Above the water level the ore occurs in irregular replacement chambers of comparatively small size, and consists of argentiferous sand carbonate, some anglesite, and residual nodules of galena. At the water level the sulphide is in greater abundance but shows some carbonate ore and a soft, black mineral, which is probably argentite. The carbonate ore is said to carry about 200 ounces of silver to the ton and the sulphides about 100 ounces to the ton.

MEGER TUNNEL.

The Meger tunnel (No. 18, Pl. X, p. 64) is a crosscut about 150 feet long, running west from the head of Copper Canyon opposite the Nevada Mining Co.'s shaft. At the face of the tunnel there are two narrow fractures in siliceous shale and quartzite, which strike N. 10° W. and dip 50° W. parallel to the bedding. In some places these fissures open out to 4 or 6 inches in width and are filled with galena and sand carbonate with some pyrite and sphalerite. The ore has been followed south for about 200 feet and stoped in several places. Open pits on the croppings have also yielded some carbonate ores.

VIRGIN AND SUPERIOR LODES.

The Virgin and Superior lodes (No. 31, Pl. X, p. 64) are on the east side of Copper Canyon, about 2½ miles south of Galena. Copper ores were known to occur at this place as early as 1866, but were not worked to any great extent previous to 1870, though the

Battle Mountain Mining Co., an English concern, obtained possession of the property in 1869. Considerable oxidized copper ore was shipped to Swansea, Wales, by this company until the late eighties. The property was idle for a number of years, and in 1907 was acquired by the Glasgow Western Mining Co., which has made some production since that time and in the summer of 1912 were installing a copper-leaching plant at the mine and reopening the shaft on the Virgin vein.

This deposit of copper ores lies in altered light-colored quartzites and lime shales below a belt of massive dark quartzite and about 300 feet east of a mass of intrusive granite porphyry. For 500 feet east of the porphyry the country rock is much broken by a series of fractures, which strike N. 10°–20° E. and dip 45°–60° W. nearly parallel to the bedding of the sedimentary rocks. The Virgin vein is the most prominent fracture on th ewest side of the zone. It ranges from 4 to 10 feet in width, but at one place opens out to a maximum width of 40 feet. It is developed by an incline said to be 600 feet deep, which was caved at the 200-foot level. The 113-foot and 188-foot levels were open in August, 1912. From the surface to the 113-foot level the ore occurred in irregular stringers and lenses throughout the width of the fractured quartzite. It consists of masses of copper carbonates, cuprite and iron oxide, with some native copper on the joint planes. 'The quartzite between these masses is all impregnated with chrysocolla and malachite, so that the whole mass is stained green. At the 188-foot level the chalcocite zone has been cut and the ore consists of chalcopyrite largely altered to chalcocite with a small amount of malachite on the joint planes. The quartzite between the masses of ore minerals is not impregnated with copper carbonates.

From the south drift at the 188-foot level, about 200 feet from the shaft, there is a crosscut east to the Superior vein, which dips 45° W. and strikes more nearly north, as if it should join the Virgin vein north of the shaft and at depth. The Superior vein resembles the Virgin, except that in some places in it at the 188-foot level there are masses of dark garnet with galena, pyrite, sphalerite, and chalcopyrite.

On the surface the Virgin and Superior veins are about 150 feet apart, and all the broken quartzite between them is more or less stained with oxidized copper ores. In this mass there are some northerly striking vertical fissures filled with limonitic material, which are said to carry high gold values.

ELDORADO PROPERTY.

The Eldorado property, consisting of three claims, is in the low hills on the east side of the mouth of Willow Canyon (No. 10, Pl. X,

p. 64). The principal development work is a 60-foot vertical shaft having a short drift to the north at the bottom. The country rock is white quartzite with a distinct bedding that strikes north and dips 50° W. The ore makes along a fracture which strikes N. 60° E. and dips 60° N. The fracture zone is about 3 feet wide and shows a little copper stain on the joints of the breccia for a few inches below the hanging wall. At the bottom of the shaft a small amount of pyrite and much less galena are seen in the crushed quartzite.

ELKO-LANDER.

The Elko-Lander vein (No. 11, Pl. X, p. 64) is on the west side of the summit of the ridge between Willow Creek and Copper Canyon, about 2 miles south-southwest of Galena. It is developed by an open cut and by a tunnel, which was locked at the time of visit. The country rock is a dense white quartzite, the rather indistinct bedding of which dips 80° W. and strikes N. 10° W. The vein, which lies parallel to the bedding of the quartzite, is about 2 feet wide and consists of honeycombed iron-stained white quartz, showing remnants of pyrite and in places specks of galena. This ore is said to carry about $40 a ton in gold and silver.

PLACER MINES.

Placer gold has been recovered from several localities in the Battle Mountains, and in 1912 dry-washing operations were underway in Box Canyon, at the mouth of Copper Canyon, and near Bannock. In the north fork of Licking Creek, both above and below the Miss Nevada workings (No. 20, Pl. X, p. 64), some placer gold was recovered several years ago.

Most of the placer gulches at the south end of the mountains are narrow and V-shaped. The gravels are all angular and roughly stratified by stream action. The principal pay streak is on the bedrock, but fair values are found throughout the gravels to the surface. The gold occurs in rather fine particles, though some good-sized nuggets have been found. Limonite, quartz, and in places pyrite are attached to the gold. All the gold is of very local origin and has at most traveled less than a mile from its source. In Box Canyon the side gulches from the ridge between Copper and Box canyons are the richest, and only the lower part of Copper Canyon, which drains off the same hill on which the Tomboy (No. 28, Pl. X) is located, has proved to be productive ground. Placer gold has been found in many of the canyons on the east and south sides of the Battle Mountains. In 1913 and 1914 there was considerable placer work done, and a number of articles have appeared in the mining journals relative to the deposits.

SKOOKUM DISTRICT, LANDER COUNTY, NEV.

LOCATION AND GENERAL GEOLOGY.

The Skookum (No. 8, Pl. I, p. 18), in southern Lander County, was one of the "one season" mushroom mining districts, which are from time to time heralded as the latest wonder of Nevada. The district covers the south end of a low group of hills east of the Reese River valley, 7 miles west of Austin. The Nevada Central Railroad skirts the eastern base of these hills to Ledlie station, where it turns east to begin the climb to Austin. This group of hills is shown in the southeast corner of Plate V of the Atlas of the Fortieth Parallel Survey[1] under the name of Jacobs Promontory.

Jacobs Promontory, according to Hague and Emmons,[2] is "made up of highly metamorphosed dark-blue quartzites with cherty seams, * * * having a general north-south strike but whose structure is very obscure." They go on to say that volcanic rocks are found over the greater part of the surface. An augite andesite flow covers a large area at the north end of the mountains and also a considerable portion of the southern hills. This andesite Hague and Emmons considered to be older than the rhyolite of the western part of the mountains, which is distinctly older than the dark fine-grained basalt that caps the summits of the hills. The quartzites are mapped as belonging to the Weber quartzite of the Carboniferous. These quartzites resemble the formations at New York Canyon, in the Reese River district, which are thought to be older than Carboniferous, and may be Silurian or Cambrian.

ECONOMIC GEOLOGY.

HISTORY OF THE DISTRICT.

The discovery of float from this district is said to have been made by an Indian in the autumn of 1907, on the present Gweenah property. The Indian sold his find to the Lemaire brothers, of Battle Mountain, who began operations in the spring of 1908. In February, 1908, James Watt, of Austin, located the property that is now the Watt mine. The float from these veins is said to have been exceptionally rich silver ore carrying some gold. When the discoveries became known a stampede began and the country for miles was staked. Two tent towns, Skookum and Gweenah, about 1½ miles apart, grew up in a night, and by midsummer were occupied by about 200 people. Those who did not own ground leased near the original

---

[1] King, Clarence, Geographical and topographical atlas: U. S. Geol. Expl. 40th Par., Pl. V, 1876.

[2] Hague, Arnold, and Emmons, S. F., Descriptive geology: U. S. Geol. Expl. 40th Par., vol. 2, pp. 641–643, 1877.

Skookum and Gweenah discoveries. W. C. Higgins[1] gives a description of the camp during this boom, which lasted till the autumn of 1908. In the autumn of 1912, with the exception of the dumps and the whim house at the Gweenah, the hills were as bare as before the discovery, although the annual assessment work is kept up at the Lemaire and Watt mines.

CHARACTER OF THE ORE DEPOSITS.

General features.—The veins of the district occur in the dark fine-grained quartzites at the south and southwest sides of the hills. They are small and are filled with white, coarsely crystalline quartz, evidently deposited in open fissures, though some of the veins have been brecciated since their formation. The ore minerals are tetrahedrite and small quantities of pyrite intergrown with the quartz. These minerals are not very abundant in any of the ore seen on the dumps and probably constituted less than 2 per cent of the best ore taken out. The veins are not easily traced on the surface and trenching was the usual means of prospecting. The surface ores are somewhat iron stained and in many places a thin film of copper silicate coats the quartz. Some of the chrysocolla is bluish or purplish and the thin coatings were mistaken by the miners for the silver chloride, horn silver. All the tetrahedrite carries silver and iron and is similar to the ore found at Austin.

The vein materials are comparable to the antimonial silver deposits in southern Humboldt County, described by Ransome.[2] He states that "beyond the fact that the ores are post-Triassic, their age is unknown, but it is probable that they are pre-Tertiary and were deposited during or after the post-Jurassic intrusions and folding that affected the whole Great Basin region and the Sierra Nevada."

There are no reliable records of the production of the district, but the total output is probably quite small. The ore so far shipped has been hand sorted and as a consequence was only the highest grade. In this ore there were about 750 ounces of silver for each ounce of gold.

Skookum property.—The Skookum property of George Watt is 1½ miles west-northwest of Ledlie station, at the south end of Jacobs Promontory. The vein strikes north and at the surface dips 60° W., but at the 100-foot level it dips 70°–85° W. It occurs near the center of a fault breccia consisting of small angular fragments of black quartzite and about 20 feet east of a rock that appears to be a dike of rhyolite porphyry. The vein ranges from 4 to 12 inches in width

[1] Higgins, W. C., Skookum, Nevada's new chloride camp: Salt Lake Mining Review, vol. 10, No. 2, pp. 17–21, 1908.

[2] Ransome, F. L., Mining districts in Humboldt County, Nev.: U. S. Geol. Survey Bull. 414, pp. 42–43, 69, 1909.

and consists of white quartz carrying argentiferous tetrahedrite that is slightly iron stained. In places thin films of copper silicate coat the quartz. Postmineral movement has brecciated the vein quartz and produced gouge on both walls. The quartz breccia is only partly cemented and on the fragments there are very small quartz needles. At the 100-foot level a minor zone of movement, which strikes N. 70° W. and dips 70° S., has displaced the north side of the vein 1 foot to the east. The main shaft is sunk 100 feet on the vein, at which level there are drifts north and south totaling 170 feet. There is also a crosscut 50 feet east of the vein along the south side of the eastward-striking fault. Above the 50-foot level the vein has been stoped to the surface for about 20 feet south of the shaft. The Skookum has been traced northward by several shallow shafts and cuts for about 500 feet.

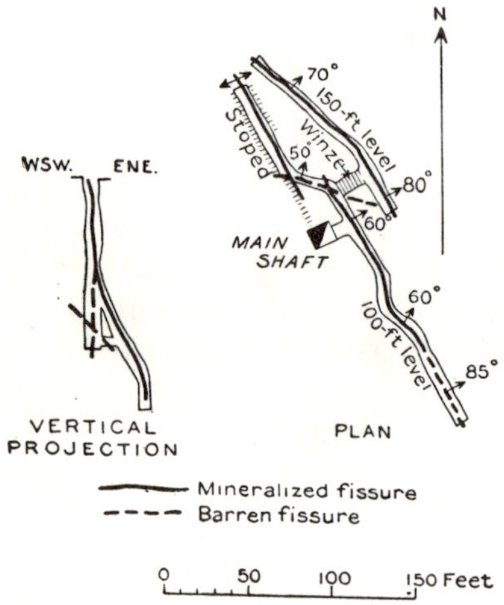

FIGURE 2.—Plan and elevation of the Gweenah mine, Skookum district, Lander County, Nev.

Gweenah property.—The Gweenah property belongs to the Lemaire brothers, of Battle Mountain. It is located about 1½ miles north of the Skookum in the southwestern part of the hills. The collar of the main shaft stands at a barometric elevation of 6,750 feet. The chief development consists of a 100-foot vertical shaft, which has about 240 feet of drifting at the first level, and a 50-foot winze below this level, which has 120 feet of drifts from its bottom. The shaft is equipped with a whim, and the winze in 1912 was so connected with the shaft that it was possible to raise all material from the 150-foot level by horsepower.

The vein strikes N. 28° W. and for the upper 60 feet is nearly vertical. (See fig. 2.) Below that depth it flattens to about 60° at the 100-foot level, but again straightens to 70° or 80° at the bottom of the winze. The dip, where there is any, is east-northeast. The apparent flattening of the vein is probably due to its displacement by a fault zone that strikes N. 75° W. and dips 50° N. This fault is shown north of the shaft on the 100-foot level and in the open stope which extends back to the shaft about 40 feet above the first level. The part of the vein north of the fault has been moved about 30 feet to the northwest along the fault plane. In the shaft there is an unmineralized fracture, much smaller than the vein fissure, which continues at least to the bottom of the shaft and is nearly vertical.

The vein has been brecciated and some gouge produced by subsequent movement, and it is possible that the vertical fracture in the shaft is to be correlated with this postmineral fracturing.

The vein ranges from 2 inches to 4 feet in width, and consists of white vitreous quartz easily distinguished from the dark wall rock. All the ore contains some tetrahedrite and is iron stained. It also generally shows thin coatings of chrysocolla, which has been mistaken for silver chloride. The ore is not evenly distributed, but occurs in shoots. The largest and, according to report, the richest shoot was that found north of the shaft at the 100-foot level, above the postmineral fault. This shoot has been stoped for about 100 feet along the vein, which is 2 to 3 feet wide, and for a height of 20 to 60 feet.

The Gweenah vein has been opened in several places southeast of the main shaft by various lessees. There are apparently several displacements of the vein, as the different shafts do not line up as they should on an undisturbed vein. There is also the possibility that there is more than one vein on the ground. In fact, it is pretty clearly demonstrated either that the Gweenah vein branches or that there are at least two nearly parallel veins about 100 feet apart.

Other properties.—There are many openings in the hills in the vicinity of the above-described properties, some of which show small veins quite similar in mode of occurrence and in mineralization to those described. Most of the work consists of pits and open cuts, descriptions of which are unnecessary.

FUTURE OF THE DISTRICT.

In the hills covered by the Skookum district there has unquestionably been some mineralization. From the development work it appears that, though some of the ores are rich, they do not occur in large veins or in very large shoots in any one vein. The fact must be borne in mind that mining small veins costs as much as working large ones, and that the returns are proportionately smaller to the cubic yard of material stoped where much waste has to be broken. It is probable that the richer ores occur at or near the surface, and that the value will decrease in marked degree with depth.

REESE RIVER DISTRICT, LANDER COUNTY, NEV.

LOCATION AND ACCESSIBILITY.

The Reese River or Austin mining district (No. 9, Pl. I, p. 18) occupies a belt approximately 6 miles in length from north to south, in the central part of the Toyabe Range, Lander County, Nev. The town of Austin is located at the center of the district. As originally organized on July 17, 1862, it was described[1] as "bounded on the north by an east-west line 10 miles north of the Overland telegraph line (Pony Canyon), on the east by Dry Creek, on the south by an east-west line 10 miles south of the Overland telegraph line, and on the west by Edwards Creek." As described in this report, the district covers the productive ground on the west side of the Toyabe Range from Telegraph Peak on the north to and including Crow Canyon on the south. The mines in New York and Yankee Blade canyons are 3 to 4½ miles north of Austin, and those in Crow Canyon 1 mile to 2 miles south of that town, as is shown in Plate XI.

Austin, the county seat of Lander County, is beautifully situated in Pony Canyon at an elevation of about 6,800 feet. The town is built on the hills on either side of the dry creek bottom and extends northwest up Cedar Gulch, which heads on Mount Prometheus. The streets of Austin afford a beautiful view over the Reese River valley and far to the west to the New Pass Mountains.

The narrow-gage Nevada Central Railroad connects the county seat with the main line of the Southern Pacific Railroad at Battle Mountain. Daily trips each way over the 93 miles of poorly built roadbed are usually accomplished.

TOPOGRAPHY.

The Toyabe Range is low in the vicinity of Austin and is marked by rounded hills, which culminate in Mount Prometheus, east of the town. Telegraph Peak, 7 miles north-northeast of the county seat, attains an altitude of approximately 9,500 feet. Telegraph Pass, 3 miles northeast of Austin, has a summit about 6,800 feet above sea level. On the western slope of the range south of Telegraph Peak there are several well-marked canyons named as follows, beginning at the north: New York, Yankee Blade, Emigrant, Slaughter House (just north of Lander Hill), Pony (in which Austin is built), Marshall, and Crow, at the south end of the area shown on Plate XI.

GEOLOGY.

GENERAL FEATURES.

The productive part of the Reese River district is underlain by a medium to coarse grained granitoid rock intrusive into sedimentary quartzites and shales, which are exposed on the southern flanks of Telegraph Peak in Yankee Blade Canyon, just north of the Patriot vein, and near the mouth of New York Canyon, on the Dictator ground. (See Pl. XI.) There is no question of the intrusive character of the contact, as many small offshoots from the large granitic body are seen in quartzites.

[1] County recorder's record, vol. 1.

The quartz monzonite is cut by a series of north-south basic dikes, which seem to be more abundant on the west slope of Lander Hill, west of Austin, than east of the town, and are seen in greater numbers in lower Marshall Basin than on Lander Hill.

The youngest rock in the district is a pinkish vesicular rhyolite porphyry flow which forms a thin cap rock on Mount Prometheus east-northeast of Austin and occurs along the summit of the ridge south of that peak for some distance, except where it has been removed by erosion. At the base of this flow, as pointed out by Emmons,[1] there is a thin contact phase of dark brownish-black glass.

SEDIMENTARY ROCKS.

Old sedimentary rocks are exposed in the extreme northern part of the area, as shown in Plate XI, on the Patriot claim in Yankee Blade Canyon, and on the True Blue claim in New York Canyon. At the base of the series at Yankee Blade there is a belt about 100 feet thick of reddish siliceous mica schists that are clearly metamorphosed quartzites, as in the upper part of the zone they grade into the unaltered quartzites. Between the micaceous quartzite schists and the intrusive granular rock there is a zone of varying width in which so much igneous material has been injected that it is difficult to determine whether the rock is igneous or sedimentary.

In New York Canyon the igneous rock is intrusive into dark, nearly black, limy shales, in which there are many stringers of white calcite. These shales seem to underlie the quartzites of Yankee Blade Canyon. The series was mapped as Weber quartzite[2] by the geologists of the Fortieth Parallel Survey, and described by Emmons[3] as "composed of highly metamorphosed beds, which include fissile limestone shales, more or less siliceous clay slates, and locally schistose to somewhat crystalline rocks resembling mica and hornblende schists." The sedimentary rocks forming the south base of Telegraph Peak are probably older than the Carboniferous, though no definite information as to their age was obtained during this reconnaissance.

INTRUSIVE ROCKS.

General character.—A gray, rather coarse to medium grained, granular rock underlies the major part of the area shown in Plate XI; in fact, all the claims south of the Patriot vein are in the "granite" belt. Emmons[4] called this rock a "normal granite, consisting of quartz, feldspar, and mica; the feldspars of two varieties,

---

[1] Emmons, S. F., Geology of the Toyabe Range: U. S. Geol. Expl. 40th Par., vol. 3, p. 329, 1870.

[2] Geological and topographical atlas: U. S. Geol. Expl. 40th Par., Map V, east half, 1876.

[3] Op. cit., p. 324.

[4] Idem, pp. 328–329.

a semitranslucent orthoclase and an opaque white variety, probably oligoclase; the mica a dark magnesian variety; hornblende is found as an accessory ingredient." From this description of the rock it would seem that it should be called a quartz monzonite, which, in fact, it is, judging from the thin sections of specimens collected by the writer.

In hand specimens the rock ranges from medium granular to coarsely porphyritic with pink orthoclase phenocrysts, the largest one-half inch in diameter. It appears to be largely composed of feldspars with subordinate femic minerals, and quartz is only rarely visible to the unaided eye. Under the microscope the rock is seen to be at least 60 per cent feldspar, but the quartz exceeds the femic minerals. Named in the order of decreasing abundance, the minerals composing this rock are orthoclase, some of which occurs in a micrographic intergrowth with quartz; oligoclase-andesine; quartz, strongly pleochroic greenish-brown biotite, and small quantities of green hornblende altered to calcite and epidote. Apatite and magnetite are abundant accessory minerals, and some sphene is present in a few of the slides.

An offshoot from this intrusive that cuts the quartzites about 150 feet north of the Patriot vein is a quartz diorite porphyry carrying more femic minerals than the large mass and little orthoclase. It is a fine-grained, nearly black porphyry dike about 10 feet in width.

The quartz monzonite is cut by a series of dikes that strike north or northwest and are in most places less than 5 feet wide. In only a few places are these dikes more than 15 feet in extreme width. There are two varieties. A fine-grained dark porphyry is the most common type, but in many places this type is not porphyritic, and has been bleached to a dull greenish-gray color. They are lamprophyres, but all the specimens collected are too much altered for further identification. The best-preserved rock of this type shows a somewhat glassy groundmass, in which lie abundant small much-altered crystals of feldspar and both pyroxene and amphiboles. There was also olivine in the rock, but that mineral is entirely altered to epidote. A small offshoot from a dike of this type which occurs on the south side of Pony Gulch, about opposite the International Hotel, is composed of black tourmaline; in fact, it is more like a vein than a dike. The tourmaline is accompanied by a little quartz and apatite, and all three minerals can be seen in the quartz monzonite for about 2 inches on either side of the half-inch dike.

In the Austin-Manhattan tunnel several of the lamprophyric dikes are cut; they are earlier than the northwest-southeast vein system and are all more or less pyritized, as is the wall rock adjacent to them. They are usually bleached, owing to the almost complete alteration of the rock to sericite and light-green chlorite, and to the addition of silica. The contacts of these dikes with the granite are remarkably sharp, in many places "frozen" but in some places separated by a thin clay seam. At the frozen contacts crystals of pyrite extend from the dike into the altered quartz monzonite. The alteration of the quartz monzonite is marked near the dikes, but the altered zone is narrow, ranging between 2 and 10 inches, depending on the width of the dike. The alteration is mainly a sericitization of the feldspars, the orthoclase apparently being first affected, and a bleaching of the biotite to a rather ragged looking aggregate of white mica and a grayish-yellow chloritic material. The quartz and apatite appear to be unaltered, but no magnetite remains.

Besides the dark dikes cutting the quartz monzonite there are a few light-colored siliceous dikes composed of fine-grained aplitic rocks that contain only a small amount of femic minerals. They are composed of quartz, orthoclase, and a little plagioclase.

Age of the intrusion.—The quartz monzonite and associated dikes are certainly intrusive into the sedimentary rocks that are probably older than Carboniferous. Beyond that their age is indeterminable in the Reese River district. The rocks, however, so clearly resemble the intrusive rocks farther west in the Great Basin province and in the Sierra Nevada, which have been clearly proved to be of late Cretaceous or early Tertiary age, that with little question the granular rocks of the district should be correlated with them. The dikes cutting the quartz monzonite are earlier than the first formed and major veins of Lander Hill, as is shown at several places in the Austin-Manhattan tunnel, where the quartz rhodochrosite veins cut both acidic and basic dikes.

The mineralization of the veins probably took place immediately after the intrusion of the small dikes or was accomplished by solutions during their intrusion.

RHYOLITE FLOWS.

The summit of Mount Prometheus and the eastern slope of the Toyabe Range, east and southeast of Austin, are covered with a pinkish to brownish purple vesicular rhyolitic rock of which Emmons[1] says: "A microcrystalline feldspathic paste incloses crystals of glassy feldspar and magnesian mica in large quantities, with occasional grains of smoky quartz. This flow rests on deeply weathered quartz monzonite, and there is no question that it is much younger than that rock." It is with little question referred to the

[1] Emmons, S. F., Geology of the Toyabe Range: U. S. Geol. Expl. 40th Par., vol. 3, pp. 329–330, 1870.

late Tertiary period of volcanism so characteristic of the Basin province.

ORE DEPOSITS.

HISTORY OF THE DISTRICT.

It is said that ore was first discovered in the district by accident. On May 2, 1862, while William Talcott, a "rider" of the Overland Pony Express, was descending Pony Canyon on his westward trip he noted that his horse kicked up some ore where he slipped on a steep pitch. Talcott took some samples to Virginia City, where an assay proved that they were rich silver ores. The Pony claim, now called the Red Bluff, just east of the Austin-Manhattan Mill (see Pl. X), was located by Talcott on his return trip on May 10, 1862. During the following summer a number of prospectors worked in the district, living at Clifton, a settlement located where the railroad station now stands. According to volume 1 of the records of the county recorder, the Keystone, Highland Mary, and Overland were located in July, 1862, and the Oregon, North Star, and South American in December of the same year. A meeting of the few prospectors in the district was held on May 10, 1862, but on July 17, 1862, the first regular "miners' meeting" was held at Jacobs Ranch (now Ledlie), a station on the Nevada Central Railroad, at which meeting the district was outlined, as given on a previous page, and the district rules adopted. In substance the rules were that a claim should be 200 feet long by 200 feet on either side of the vein; that location could be made either on "croppings" or on veins discovered in a tunnel; that the claim should be recorded within 10 days of location; and that if $1,000 was expended on a claim, it could not be relocated unless the original claimant formally acknowledged his abandonment of the location.

Austin became the county seat of Lander County September 2, 1863, and has retained that position till the present time. Buells Mill, of five stamps, was built in 1863 and started operations in August of that year. By 1883 there were 29 mills with 444 stamps in the district, but not all of them were working, as there was not enough ore mined to supply that many stamps. In 1866 Browne[1] estimated that the annual production since the discovery of the district had amounted to about $900,000. He says that there were 17 mills with 200 stamps at that time treating about 150 tons of ore a day at a cost of $45 a ton. The ores, he says, contained chlorides, bromides, and antimony-arsenic compounds of silver, and ran from $100 to $200 a ton. There were 36 steam hoists in operation, and the deepest work in the district was down 300 feet.

[1] Browne, J. R., Mineral resources of the States and Territories west of the Rocky Mountains, 1866, pp. 33, 84, 128, 246.

In 1867 Browne[1] says there were 6,000 locations in the district, each 500 by 2,000 feet in size. The production from July 1, 1866, to August 1, 1867, was $1,455,273.60.

In 1869 Raymond[2] reported that there were 4,000 persons in Austin, but that many mines were idle; a large number of men had gone to the White Pine district, being disappointed in the small size of the veins at Austin. At that time (1869) it was noted that the veins at the west end of Lander Hill were low grade and carried a large proportion of the base metals. In the report for 1869 Raymond gives a good description of the Mettacom mill, formerly located at Yankee Blade but now entirely demolished.

In 1870 Raymond[3] reports that the Buel North Star was the only important company working, though much leasing was done on Lander Hill. The Manhattan mill was treating the custom ores for $30 to $35 a ton and giving the miners from 80 to 85 per cent of the assay value of their ore.

In 1871 Raymond[4] reported that most of the Lander Hill veins were controlled by the Manhattan Co.

During the period from 1872 to 1877 the Manhattan Silver Mining Co. of Nevada, under the direction of A. A. Curtis, succeeded in obtaining practically all properties on Lander Hill, which were extensively worked under the leasing system. This company continued operations till August, 1887, when the Manhattan mill was finally shut down. This mill from its inception produced $19,239,032.87 worth of bullion from custom and company ores from Lander Hill, which were of too low grade to stand shipment to the smelters. The average fineness of the bullion produced was 0.800.

In 1888 the property of the Manhattan Silver Mining Co. was purchased by the Manhattan Mining & Reduction Co., who did little work and finally in 1894 sold its holdings to the Austin-Manhattan Mining Co., under J. Phelps Stokes, of New York. During the next 10 years this company acquired more ground and to develop Lander Hill drove the Clifton (Austin-Manhattan) tunnel, a 6,000-foot crosscut which has its mouth near the railroad station at the mouth of Pony Canyon. Under Mr. Stokes's management a 40-stamp concentrating mill was built at the tunnel mouth, and some ore was mined and milled. In 1904 the Nevada Mining Co. acquired the Stokes property but held it only one year. After passing through the hands of the Austin-Hannapah Mining Co. it came into the possession of the Austin-Manhattan Consolidated Mining Co. in 1906.

---

[1] Browne, J. R., Mineral resources of the States and Territories west of the Rocky Mountains, 1867, pp. 394–402.

[2] Raymond, R. W., Mining statistics of the States and Territories west of the Rocky Mountains, 1869, pp. 118, 123, 733.

[3] Idem, 1870, p. 11.

[4] Idem, 1871, p. 168.

This company has done no mining, though some of its ground has been worked under lease. In September, 1912, the consolidated company's holdings of 62 patents, 58 locations, 4 mill sites, and 160 acres of patented placer ground were held by a receiver. These properties were reported sold to H. C. Fownes, of Pittsburgh, in December, 1913.

Besides the Austin-Manhattan Consolidated holdings there are four other companies who control relatively small groups of claims in the Reese River district, and there are a few individual holdings.

The only large operator in 1912 was the Maricopa Mines Co., which was working in New York Canyon on the western end of the Patriot vein. In August, 1912, it had erected a 60-ton cyanide mill equipped with tubes for fine grinding.

PRODUCTION OF THE DISTRICT.

The total production of the Reese River district is commonly estimated at $50,000,000, though figures as high as $65,000,000 are sometimes given. Statistics of the output from the smaller mines are lacking, but the production from some of the better-known veins is tabulated below, the figures being supplied by the receiver of the Austin-Manhattan Mining Co.

Production of the most prominent veins of the Reese River district.

Sonora	$600,000	San Jose	$750,000
Farrell	3,000,000	Silver Chamber	1,500,000
Ruby	1,500,000	Witlatch	100,000
Oregon	3,000,000	Carmago	100,000
Isabella	1,500,000	Baker	1,500,000
North Star	5,000,000	Paxton	1,000,000
Independence	3,000,000	London	1,500,000
Frost	2,000,000	Savage and Diana	2,000,000
Panamint-Bowdie	9,000,000		
Union	3,500,000		40,550,000

The following table is compiled from various sources, most of the figures being taken from the reports of the Director of the Mint. As will be seen, the figures are by no means complete, but they give some idea of the fluctuations of the yearly output of the district.

Production of silver in Reese River district, 1862–1892, inclusive.

1862–1867	$4,500,000	1888	$29,987
1868–1881	20,000,000	1889	----------
1882	878,250	1890	82,577
1883	1,182,185	1891	1,579,666
1884	1,128,910	1892	97,397
1885	1,120,100		
1886	980,000		31,810,529
1887	231,457		

Since 1902 the figures of production of the mining districts of the several States have been carefully collected by the United States Geological Survey. The following table is compiled from the mineral resources reports issued yearly by the Survey:

Production of Reese River district, Lander County, Nev., 1902–1911, inclusive.

	Crude ore.	Gold.	Silver.	Copper.	Lead.	Total value.
	Tons.		*Ounces.*	*Pounds.*	*Pounds.*	
1902	50	$550				$550
1903	3,000	10,000				10,000
1904						
1905	205	1,605	22,321			15,087
1906	193	1,130	54,634	3,109	14,000	39,170
1907	586	667	27,787	2,640		19,535
1908	161	1,355	15,935	212	6,691	10,119
1909	618	3,272	9,127	1,562	19,837	9,074
1910	674	3,764	27,159	6,797	72,076	22,464
1911	3,043	21,801	101,592	1,972	29,067	77,199
	8,530	44,144	258,555	16,292	141,671	203,198

Previous to 1905 no attempt was made to obtain correct statistics of the base-metal output from Reese River district, but it is probable that there was little if any output of these metals before that date. In 1894 the Austin-Manhattan concentrating mill was built. It is possible that some lead and silver were obtained from concentrates produced at that mill, though the figures are not available. The production given above is all from ore shipped to the smelters, as no mill has been run for any length of time since 1904.

Some experimental work at the Austin-Manhattan mill in 1905 resulted in the removal of the vanners and in the installation of eight electrostatic separators, which have proved worthless for the concentration of the Reese River ores. This mill and the New Maricopa mill, in New York Canyon, are the only mills remaining in the district.

CHARACTER AND DISTRIBUTION OF THE DEPOSITS.

The ores of the Reese River mining district occur in simple veins, mostly cutting quartz monzonite, but are also found in the schistose quartzites on the lower south slope of Telegraph Peak. As Emmons[1] pointed out years ago:

> The veins are very small, and occur in series that have a generally parallel trend northwest to southeast and dip to the northeast at various angles between 15° or 20° and 60° or 70°. They vary in width from a mere seam, that may be traced with difficulty, to 2 or 3 feet; the pay streak seldom maintains for any considerable length a greater width than 2 or 3 inches, though frequently expanding for short distances to 5 or 6, and in exceptional cases measuring 18 or 20 inches.

[1] Emmons, S. F., Mining and milling at Reese River: U. S. Geol. Expl. 40th Par., vol. 3, pp. 349–393, 1870.

Innumerable veins outcrop on the southern slopes of Lander, Central, and Union hills, and several in the Yankee Blade region and on the low southern slopes of Telegraph Peak. This prevalence of outcrop on the southern slopes is accounted for by the low north-northeast dip of the veins, as most of the northern slopes are nearly parallel to the dip, and hence disclose few croppings.

From the few veins seen by the writer, supplemented by information obtained from the men familiar with the old workings and a study of the maps, it would seem that the more persistent veins are those with steeper dips. The flat veins apparently run into the steep veins with depth, and it is undoubtedly true, as pointed out by S. F. Emmons,[1] "that the number of veins or fissures cut in the deeper crosscuts and shafts, in various parts of the hill, bears a very small proportion to the number of outcrops at the surface in their immediate vicinity."

GANGUE MINERALS OF THE VEINS.

At the surface the white quartz gangue is somewhat rusty and seamed with black lines, owing to the alteration of the original rhodochrosite of the ore. The black gangue minerals at the surface are largely pyrolusite, the soft, sooty manganese oxide, though some psilomelane is present.

With depth the gangue is white and dark quartz and rhodochrosite, manganese carbonate, associated in places with a little calcite. These minerals are arranged in bands in true vein form. Next to the quartz monzonite walls, and in many localities frozen to them, there is generally a narrow band of quartz that in some places is white and in others dark colored. Inside of these bands there is deposited a layer of rhodochrosite and quartz, either predominating in different parts of the veins, and the centers of the veins are filled with white quartz. This simple banding in some places is repeated, particularly in the wider portions of the veins, but the succession, as a rule, is that given above, the expanding width of the bands forming the wider veins.

ORE MINERALS.

The ore minerals mostly occur in bands inside the outer quartz, but in a few places in the center of the vein. They are chiefly associated with white quartz, and as far as noted do not occur in the carbonate gangue, though bands of sulphides are present on both sides of the rhodochrosite streaks. S. F. Emmons says:[1]

> The ores generally consist of rich silver-bearing minerals, comprising pyrargyrite, proustite, stephanite, polybasite, tetrahedrite, argentiferous galena, zinc blende, which is believed to carry silver, copper pyrites, and iron pyrites.

[1] Op. cit., p. 351.

From the studies of ores collected by the writer it would seem that the original minerals were deposited in the following order: Pyrite and chalcopyrite, arsenopyrite, galena, sphalerite, wurtzite, and lastly tetrahedrite, which is rich in silver. The pyrite, chalcopyrite, and arsenopyrite, are more or less intergrown and are surrounded by the lead and. zinc minerals. Tetrahedrite is not surrounded by the other constituents, but forms about the other minerals and in cracks in them. All these sulphides, with the exception of pyrite, appear to be silver bearing, though there is some question as to the silver content of the originally deposited sphalerite.

ALTERATION OF THE QUARTZ MONZONITE BY THE VEIN-FORMING SOLUTIONS.

Near the veins the walls are softened and bleached. This alteration, though very distinct, is quite local, few of the bands of altered rock exceeding 10 inches in width, and most of them being between 1 and 3 inches. Where a vein split, inclosing a "horse" of the wall rock, all the material in the "horse" as a rule is altered. In a few places, where a number of small veins occurred in a space of a few feet, all the adjacent rock was altered.

Where the veins are frozen to the walls the quartz monzonite for about one-half inch from the veins has been somewhat silicified. The principal alteration, however, is sericitization. The quartz and apatite of the quartz monzonite appear to remain unchanged, even at the immediate contact with the veins. All the feldspars have been altered to masses of sericite and quartz. The biotite is bleached to a white sericitic mica, though a little chloritic material is developed in some of it. No magnetite remains near the veins, but pyrite is present in small crystals scattered through the soft altered rock. This soft bleached rock is cut by minute veinlets of a carbonate mineral that seems to be calcite, though it may possibly be rhodochrosite.

Aside from the pyrite the altered wall rock generally contains no metallic minerals. It is said, however, that along the main shoot of the Oregon vein some ruby silver was found in joints in the wall rock adjacent to the vein.

FAULTING OF THE VEINS.

All the veins of the Reese River district from New York Canyon to Marshall Basin have been displaced by a series of approximately parallel faults, which strike nearly north and south and dip to the west at medium or fairly steep angles. The movement along these faults seems to have been normal, dropping the part west of the fault, so that the veins are displaced to the south. The amount of movement ranges between 10 and 70 feet in those "breaks" seen by the writer. S. F. Emmons[1] says:

[1] Op. cit., p. 355.

The extent of the faulting by these north and south veins may vary from 30 or 40 feet to perhaps several hundred. * * * The cross veins [faults] are usually very narrow fissures, carrying little else than a seam of clay, and, so far as known to the writer, are not ore bearing to any noteworthy extent. * * * Besides these more important faults * * * there are numerous slighter faults or breaks * * * often repeated at short intervals.

The faulting of the veins apparently took place, at least in part, very shortly after their formation, for in the Austin-Manhattan tunnel some of the basic dikes occupy fractures similar in strike and dip to the main faults of the region, and the basic dikes are thought to be the end products of the quartz monzonite intrusion.

The presence of these faults has caused endless litigation over the ownership of the Lander Hill veins, for when a fault was reached the miners started a crosscut to pick up the lost veins and worked the first vein they came to that carried good ore. The veins are so variable along their strike, both in width and mineralization, that it is next to impossible to say what vein has been cut without following it up to the surface or to known workings.

LOCATION OF THE ORE SHOOTS.

The Austin veins are not continuously mineralized along their entire length. The ores occur in distinct shoots ranging from a few feet to about 300 feet in maximum length, but averaging between 150 and 200 feet. The shoots seem to be located near the junctions of the veins with the northward striking faults discussed above and as a consequence appear to pitch to the northwest on the dip. The above facts have not been definitely established, but seem to be justified from a study of the available mine maps. It is thought that the original mineralization of the veins was not uniform throughout, and that the faults having a northerly strike caused openings to be formed along the strike of the veins which permitted descending waters to secondarily enrich the ores in their vicinity.

The richest shoots of ore occurred in the central part of Lander Hill in an area extending for about a mile in a northerly direction and about half a mile wide. In this locality there are a large number of small northward-striking faults and a few large displacements. The ore shoots of the Patriot vein in the Yankee Blade district also seem to be closely related to a series of faults having a northerly strike.

On the western end of Lander Hill, east of a north-south line through Diana claim (see Pl. XI), the more productive veins have a north-northwest strike and stand nearly vertical or dip steeply to the west. They are characterized by a much greater proportion of the sulphides of the base metals, particularly galena, sphalerite, and chalcopyrite, and the silver content of the ore is said to be much lower than that in the central Lander Hill ores.

CHARACTER AND MINERALIZATION OF THE SHOOTS.

In most of the ore shoots there is a thin clay parting on the hanging wall of the veins, and in a number of them there are narrow bands of crushed quartz and rhodochrosite parallel to the walls and rich in secondary silver minerals.

The level of ground water on Lander Hill, where most of the productive mines of the district were located (see Pl. XII), is said to have been reached at a depth of 60 feet. It is now much lower and most of the mines are dry to the level of the Austin-Manhattan tunnel, which cuts the Frost shaft at a depth of 660 feet.

Above the original water level the ores carried "horn silver and the chlorides, oxides, and some salts of the associated metals * * * locally known under the general term of 'chloride.'"[1] There is said to have been a barren zone in practically all the veins for about 30 feet immediately below the old water level.

Below a depth of 90 feet and, from all that can be learned, to the greatest depth obtained in any of the mines, the ore minerals were galena, sphalerite, tetrahedrite, arsenopyrite, and chalcopyrite, together with more or less argentite, proustite, pyrargyrite, stephanite, and polybasite.

All these minerals have been identified by the writer from ores taken from the ore shoots to the depth of the Austin-Manhattan tunnel, though at that depth the rich silver sulphantimonide minerals seem to be less abundant than in ores nearer the surface.

Argentite, Ag_2S, is the least common of any of the secondary silver minerals. In fact it was found only in the ores from the few stopes comparatively near the old water level which could be entered. Stephanite, Ag_5SbS_4, does not appear to be abundant, but polybasite, $(AgCu)_9SbS_6$, is often seen as well-developed crystals in cavities lined with drusy quartz and in specks through the richer ore. Proustite, Ag_3AsS_3 (light ruby silver), was noted in thin films on fracture planes in ore from the tunnel level of Frost shaft, 660 feet vertically below the surface. Pyrargyrite, Ag_3SbS_3 (dark ruby silver), was found at a number of places apparently surrounding and mixed with tetrahedrite, which it seemed to be replacing. Secondary pyrite was noted on drusy quartz in fractures in ore at the Maricopa mine.

---

[1] Emmons, S. F., Mining and milling at Reese River: U. S. Geol. Expl. 40th Par., vol. 3, p. 351, 1870.

The facts above stated accord with the occurrence of the secondary silver minerals, which has been so well summarized by W. H. Emmons,[1] who says:

> Although there is no invariable rule * * * the chloride is most abundantly developed above the argentite ore; the antimony and arsenic sulphosalts are found with and below the argentite.

He considers that pyrargyrite, which is abundant at Austin, is "probably the most abundant" secondary silver mineral. The rich silver minerals of Austin seem to have been precipitated around the original sulphides by their reducing action, though in one specimen of ore from the Maricopa mine there was a suggestion that dark ruby silver replaced tetrahedrite. Taylor[2] notes that pyrargyrite probably metasomatically replaces tetrahedrite in the Austin ores. Polybasite was not definitely determined in association with the original sulphides, but it is of common occurrence as fillings in quartz-lined druses, particularly at the Maricopa tunnel. Bastin[3] has shown that this mineral metasomatically replaces galena and quartz in ores from the Central City quadrangle, Colo.

VALUE OF THE ORES.

In the early days of mining at Austin extremely rich chloride ores were extracted that ran as high as $2,000 a ton. Below the water level the ores are said to have carried between $175 and $300 a ton in silver, averaging over $200 for a period of 20 years. To be sure, the total value of the ore was greater, but the silver was all that was saved in the chlorination mills of the camp. During the period from 1902 to 1911 there was mined 8,530 tons of ore, having a total value of $203,198, or $23.83 a ton.

DEVELOPMENT OF THE VEINS.

The veins on Lander Hill were first developed by inclines and drifts. Later vertical shafts were sunk and crosscuts made at the various levels to the veins, which were then followed. The maps in the possession of the Austin-Manhattan Consolidated Co. show miles of underground development, very little of which is now open. The veins were narrow and the stopes as a consequence were made small in order that as little waste rock as possible need be hoisted. Many pillars of good size were left, if the maps are correct, to support the flat-lying roofs. The numerous faults made it necessary to drive

[1] Emmons, W. H., The enrichment of sulphide ores: U. S. Geol. Survey Bull. 529, pp. 118–125, 1913.

[2] Taylor, N. B., A study of ores from Austin: Univ. School of Mines Quart., vol. 34, pp. 32–39, 1912.

[3] Bastin, E. S., Metasomatism in sulphide enrichment: Econ. Geology, vol. 8, pp. 51–63, 1913.

LANDER HILL, REESE RIVER MINING DISTRICT, AUSTIN, LANDER COUNTY, NEV.

Looking north-northeast from Marshall basin road at summit of divide south of Austin. 1, Lander shaft; 2, Curtis shaft; 3, Frost shaft; 4, North Star incline; 5, South American incline; 6, Paxton incline; 7, Isabella incline; 8, Magnolia incline; 9, Bowman incline; 10, Manhattan mill site; 11, Union shaft; 12, Great Eastern shaft.

many thousand feet of crosscuts through barren ground. The nature of the faults was discovered at an early date and the miners had little hesitancy in starting crosscuts on each level with the assurance that they would pick up their vein. If they were driving west on a vein which was faulted, they crossed the fault and turned to the south with their crosscut, and if they were working east on the vein they turned north beyond the fault, as shown in figure 3.

Most of the dumps on Lander Hill, shown in Plate XII, are at the mouths of inclines from 150 to 200 feet deep, but several of the larger dumps are at inclines from 500 to 1,100 feet in depth, and the largest dumps are at the vertical shafts, of which there are eight over 500 feet deep.

ACCESSIBLE WORKINGS.

It was most unfortunate that in August, 1912, most of the old mines were inaccessible on account of long idleness. From an inspection of the mine maps some data were obtained, but not as much information could be gathered from them as would have been possible if the miles of underground workings had been still open. The workings that could be entered by the writer were the Hardy incline in Crow Canyon, a new 100-foot shaft on the Jackpot vein at the west end of Union Hill, the upper part of the Sobe incline on the Panimint vein on Central Hill, the Austin-Manhattan tunnel under Central and Lander hills, the upper levels of the Isabella and Magnolia inclines on Lander Hill, the Moss incline at the west end of Lander Hill, and the Maricopa and Morris & Cable tunnels and the Watt incline in New York Canyon.

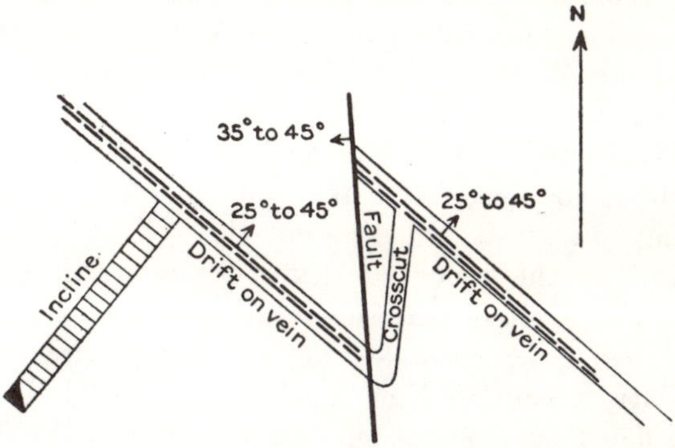

FIGURE 3.—Generalized sketch showing direction of crosscuts where veins are cut off by faults on Lander Hill, near Austin, Lander County, Nev.

Watt mine.—The Austin Goldfield Mining Co. owns the Cambrian claims, shown at the north end of Plate XI (p. 96). The main development on the Cambrian or Todd claim on the north side of New York Canyon, at the east end of the group, is a 300-foot incline shaft with three levels. It is more often spoken of as the Watt mine. The vein strikes N. 45°–50° W. and dips 25°–45° NE. in dark quartzites lying about parallel to an indistinct bedding. The vein ranges from 4 to 12 inches in width, consisting of white quartz carrying tetrahedrite, pyrite, chalcopyrite, and some sphalerite and galena.

In many places it splits, surrounding horses of black crushed quartzite that carries a large amount of carbonaceous material and crushed pyrite and somewhat resembles graphite. Below the vein the quartzite is crushed and contains disseminated pyrite. The main ore shoot occurs at the shaft, being about 120 feet long on the first level, 80 feet on the second level, and about 40 feet on the third level. On the second level, 200 feet east of the shaft, the vein is offset 12 feet by a fault that strikes north-northeast and dips northwest at low angles along which the movement was normal. Near this fault there is a shoot of ore that has been stoped to some extent. The ore being mined in 1912 is said to average about $126 a ton, of which only 80 cents was gold.

Morris & Cable tunnel.—The Morris & Cable tunnel on the Patriot vein runs east from New York Canyon, starting southwest of the Watt shaft. The Buell crosscut also intersects the Patriot, starting south from New York Canyon at a point southeast of the Watt shaft. The Morris & Cable tunnel had been cleaned out for about 430 feet from the mouth. Two northerly striking faults having low westward dips cut the vein, the offset on one being 30 feet and on the other 20 feet. Considerable stoping has been done on shoots located near the faults, but the stopes are now caved.

Maricopa mines.—The Maricopa Mines Co. leases claims on the Patriot vein in New York Canyon. Their main development is by a 550-foot crosscut tunnel, starting on the True Blue claim (see Pl. XI, p. 96) and running north into the west side of New York Canyon. The vein is about 200 feet north of the main contact of the quartz monzonite and quartzite, though many dikes of igneous rock are exposed in the workings, the vein cutting through both rocks. The vein has been followed for about 400 feet, being cut by a fault along which there has been much crushing, but apparently only small displacement. The vein in the quartzite consists of 2 to 8 feet of black graphite-like crushed quartzite cut by white quartz stringers from a fraction of an inch to 18 inches in width and carrying abundant pyrite and arsenopyrite. The quartz in many places lies near the hanging wall, but in the wider portions of the vein small veinlets occur all through the ore zone. In the quartz monzonite dikes the vein generally narrows to a width of 8 to 10 inches and consists entirely of quartz and sulphides.

The richer ores consist of quartz and a little carbonate associated with tetrahedrite, galena, sphalerite, arsenopyrite, pyrite, and chalcopyrite, the rich silver sulphantimonide minerals being secondary products in fractures in the ore which are sometimes completely filled by but more often lined with white quartz crystals, on which the secondary sulphide crystals are perched.

The ore as it is mined, taking in all the material between the walls, both quartz and crushed country rock, is said to average 25 to 30 ounces of silver and 80 cents to $3 in gold to the ton.

Isabella and Magnolia inclines.—The Isabella and Magnolia inclines on Lander Hill (see Pl. XI, p. 96) apparently start on the same vein, though underground there are so many crosscuts to different veins that it is impossible to tell without adequate maps what vein is being followed by a drift. The Magnolia incline was followed to a depth of 500 feet, and a drift west, presumably on the Corelly vein, connects with a drift from a crosscut from the 527-foot level of the Isabella incline. The Corelly vein ranges from a fraction of an inch to 20 inches in width, averaging about 8 or 10 inches in most places. It consists of banded white quartz and rhodochrosite associated with different quantities of silver-bearing minerals, largely tetrahedrite and galena, together with some sphalerite, pyrite, arsenopyrite, and chalcopyrite. The ore minerals occur in shoots, which have been very largely stoped. The longest shoot seen extended for 150 feet along the drift. The Isabella incline was entered to a depth of 170 feet and there a cave barred further exploration. The cave occurred at a fault, which strikes N. 20° W. and dips 25° W. and which has displaced the Isabella vein about 25 feet. From the surface to a depth of 75 feet the 18-inch vein has been stoped, but below that depth to the cave the vein is pinched and barren. On the maps of this incline a second shoot is shown at the 500-foot level.

Moss incline.—The Moss vein, north of the west end of Austin (see Pl. XI), carries much lead and zinc and is reported to carry some silver. The strike is N. 55° W. and the dip 69° NE. The vein consists of 10 inches to 3 feet of quartz frozen to the footwall and separated from the hanging wall by a tight slip plane. It is developed by a 100-foot incline in the vein with a short drift southeast on the 50-foot level.

Austin-Manhattan tunnel.—The Austin-Manhattan tunnel starts on the north side of the mouth of Pony Canyon, just east of the mill shown at the left center of Plate XI. The tunnel runs about S. 77° E. for 5,925 feet, to the bottom of the Frost shaft, at a vertical depth below the surface of 660 feet. A crosscut that strikes N. 59° E. is turned at a point 3,380 feet from the mouth. This crosscut is said to be 2,840 feet long but was caved in the spring of 1912. At a point 1,760 fect beyond the first crosscut there is some work south of the main tunnel, which was bulkheaded and could not be seen. Two veins are said to have been exposed in this drift. At the face of the main tunnel, which is 6 by 7 feet in the clear, is a crosscut 450 feet long that runs S. 32° E. toward the east end of Union Hill.

About 1,000 feet from the portal there is a 10 to 12 inch quartz vein carrying pyrite that strikes N. 65° W. and dips 15°–20° N. The quartz monzonite wall rock is bleached and softened for 2 inches on either side of the vein. Fifty feet farther this vein is completely cut off by a 2-inch quartz vein, carrying galena and sphalerite, which strikes N. 7° E. and dips 75° E.

About 1,700 feet from the portal a quartz rhodochrosite vein ranging from 2 to 10 inches in width enters the north side of the tunnel. This vein strikes N. 51° W. and dips 55° N. At a point 20 feet east it runs into a 12-foot dike of fine-grained light-gray quartz monzonite porphyry that carries an abundance of disseminated pyrite. This dike strikes N. 12° E. and dips 85° E. In the center of this dike there is a 2-foot barren quartz vein which completely cuts off the quartz rhodochrosite vein.

There are several other examples of both systems of veins in the Austin-Manhattan tunnel. The northerly striking, usually barren system is best shown, as the tunnel runs nearly parallel to the productive vein system of Lander Hill.

Sobe incline.—The Panimint vein was seen in the upper 80 feet of the Sobe incline, located near the Bowman incline on the Spokane ground. (See Pl. XI, p. 96.) The vein ranges from a close fracture to 3 feet in width in the 300 feet of drifts on the 80-foot level. It strikes about N. 45° W. and dips 35° N. West of the incline for 70 feet the vein, apparently about 10 inches wide, was stoped to the surface. Beyond the stopes it is pinched to a tight barren fracture. About 100 feet east of the incline the vein splits into two 6-inch quartz rhodochrosite stringers, separated by 2 feet of altered quartz monzonite. Near the top of a 35-foot stope the two stringers unite into a 12-inch quartz vein carrying some sulphides.

Jackpot vein.—The Jackpot vein at the west end of Union Hill, about one-half mile south of Pony Canyon (see Pl. XI), strikes N. 19° W. and dips 80° E. It is developed by a 400-foot vertical shaft, which could not be entered, and by a 100-foot shaft recently sunk about 300 feet north of the main shaft. In the new shaft, which has no ladders and is equipped with a hand windlass, the vein is 18 inches wide. It consists of white quartz carrying pyrite, arsenopyrite, chalcopyrite, tetrahedrite, and some galena and sphalerite, which occur in irregular stringers and bunches throughout the quartz. The mill ore is said to average 29 ounces of silver and 0.24 ounce of gold to the ton, but the hand-sorted ore that was shipped in 1911 averaged about 100 ounces of silver and 1 ounce of gold to the ton.

Hardy vein.—The Hardy vein in Crow Canyon, at the extreme south end of the district (see Pl. XI, p. 96), strikes N. 42° E. and dips 70° SE. It is developed by a 30-foot incline, at the bottom of which there is an 80-foot drift northeast on the vein. The vein ranges from 4 to 6 inches in width and consists of iron-stained quartz. The quartz seam cuts irregularly through a 4-foot zone of somewhat crushed and completely softened and altered quartz monzonite, being generally near the footwall of the zone. Aside from a small amount of pyrite, galena, and sphalerite, no sulphides were seen in the porous rusty ore, which is said to carry about $30 a ton, 24 per cent of the value being in gold and the remainder in silver.

Veins on Midas Flat.—The veins on Midas Flat south of New York Canyon, in the lower part of Telegraph Canyon, have yielded some very high grade ores. It is said the silver was largely carried as "chloride" and argentite and occurred in relatively small pockets. None of the veins were being worked in 1912. As far as could be seen from an inspection of the dumps, the ores are identical with those found in the upper levels of the Lander Hill veins. It is reported that at a depth ranging from 350 feet at the east side of Midas Flat to 150 feet at the west side of the productive ground the veins were all cut off by a very flat fault dipping west-southwest. The veins strike from N. 20 W. to due north, corresponding with those at the west end of Lander Hill rather than with the rich veins in the silver area of Lander Hill and Yankee Blade. It therefore seems rather questionable to assume, as has been done by the local miners, that the Midas Flat veins are the upper portions of the Yankee Blade veins and have been faulted into their present position.

FUTURE OF THE DISTRICT.

In view of the fact that in the palmy days of Austin ores of less value than $75 a ton were not usually hoisted, it is probable that considerable medium-grade ore was left in the abandoned mine workings as pillars and stope fillings. It is certain that there is also a fair quantity of low-grade ore in the many dumps on Union, Central, and Lander hills, which, if it could be sorted, would pay to mill under the metallurgic processes now used.

It is a question how much of the stope filling and ore left as pillars in the old workings could now be hoisted profitably, on account of the condition of the long-abandoned drifts.

The old mines have been worked to a vertical depth between 300 and 500 feet, though on the dip the veins have been stoped in some places from 1,100 to 1,200 feet below the croppings. It has been shown by the Austin-Manhattan tunnel, which is below the level of most of the old development on Lander Hill, that the veins still persist and are mineralized at that depth. It seems probable, and in fact is fairly well shown, that the ores at that depth are not as valuable as they were nearer the surface. The smallness of the veins is also to be taken into account, as the mining of narrow veins requires a large expenditure for blasting and moving barren rock. However, it seems entirely possible that with careful management and good mining a company controlling a large amount of ground (such as the Austin-Manhattan Consolidated) might be able to bring Austin again to the fore as a producer of silver.

TOYABE RANGE, SOUTH OF REESE RIVER DISTRICT, INCLUDING THE WASHINGTON AND KINGSTON DISTRICTS, LANDER COUNTY, NEV.

LOCATION AND ACCESSIBILITY.

The mines described in this section of the report are located on both sides of the Toyabe Range, largely in Lander County, south of and including the mines in Big Creek on the west side and Birch or Geneva Creek on the east side of the mountains. The area embracing the mines on these two creeks is ordinarily spoken of as the Reese River district, but the mines are so close to the south line of the recorded district that it is thought best to include their descriptions with those of the mines in Kingston (No. 11, Pl. I, p. 18), Washington, and San Juan canyons (No. 10, Pl. I). The Kingston canyon is occupied by an eastward-draining stream and the Washington and San Juan by westward-flowing creeks. The mines are all tributary to Austin. Most of the mines shown on Plate XIII are located within the Toiyabe National Forest.

Reese River valley, west of the range, is well watered and a great part of the land is under cultivation, hay and grain being the principal crops. Smoky Valley, on the eastern side, is practically all desert except for small irrigated areas near the foot of the mountains, where the creeks carry a sufficient volume of water to flow out over the Quaternary wash. Stock raising is the principal industry in this part of Lander County at present, mining being secondary since the decline of the mines at Austin. Annual assessment work constitutes the greater part of the development of many of the claims.

TOPOGRAPHY.

The Toyabe (Indian name for mountains) is a long, narrow range, which trends about N. 25° E. It rises approximately 6,000 feet above the broad valleys on either side and is scarcely 10 miles wide at its base at the widest part. The short, steep canyons appear to have quite gentle gradients as compared with their rugged walls. The eastern side of the range in this vicinity seems to be much steeper than the western flank, but, as Emmons[1] has pointed out—

> A closer examination of its topography discloses a double-ridge system [see Pl. XIII], which prevails through the greater part of its extent, giving rise to a

[1] Emmons, S. F., Mining industry: U. S. Geol. Expl. 40th Par., vol. 3, pp. 320–321, 1870.

series of interior longitudinal basins. * * * Birch Creek and Kingston Creek break through the eastern ridge, while only Big Creek flows to the west.

The following elevations, given by Emmons,[1] agree closely with those taken by the writer:

Barometric heights in the Toyabe Range, Lander County, Nev.

	Feet.		Feet.
Summit of stage road, east Austin	7,512	Big Creek Peak	10,265
Geneva Peak	10,994	Kingston	6,247
Big Creek Canyon mouth	6,400	Bunker Hill, south peak	11,602
Big Creek Pass	8,922	Bunker Hill, north peak	11,735
		San Juan	8,075

GEOLOGY.

Rocks of three distinct types are exposed in this portion of the Toyabe Range. The oldest are the sediments, now much metamorphosed, that make up the largest part of the mountains. A granular igneous rock intruded into this sedimentary series forms the second type. The somewhat glassy lava flows capping both the sediments and the intrusive granitic rock constitute the third type. Emmons's[2] mapping of the larger units, as shown on Plate XIII, gives a more accurate idea of their distribution than the writer was able to determine in his hasty reconnaissance.

PALEOZOIC SEDIMENTARY ROCKS.

Quartzite.—On the crest of the range east of San Juan there is a belt of white dense quartzite, weathering red, that lies between granodiorite on the east and slates on the west. Interbedded with this quartzite are some thin beds of white limestone. Emmons[3] states that these beds are several thousand feet in thickness, and "may belong to the Devonian series." Spurr[4] thinks that these quartzites may be equivalent to the Cambrian quartzites at Eureka, Nev., described by Curtis[5] and by Hague.[6]

Slates.—The central mass of the portion of the Toyabe Range under discussion is composed of dark thin-bedded clay and lime shales, now largely metamorphosed into siliceous and calcareous slates and to contorted schists of a prevailingly dark-gray to black color. Interbedded with these black shales are reddish-purple to green slates and lenses of dark crystalline limestone. This series, estimated by Emmons[7] to be 7,000 feet thick, overlies the quartzite

[1] Emmons, S. F., Mining industry: U. S. Geol. Expl. 40th Par., vol. 3, p. 347, 1870.

[2] Idem, Pl. XIII.

[3] Idem, p. 324.

[4] Spurr, J. E., Descriptive geology of Nevada south of the fortieth parallel and adjacent portions of California: U. S. Geol. Survey Bull. 208, pp. 94–95, 1903.

[5] Curtis, J. S., Silver-lead deposits of Eureka, Nev.: U. S. Geol. Survey Mon. 7, 1884.

[6] Hague, Arnold, Geology of the Eureka district, Nev., with an atlas: U. S. Geol. Survey Mon. 20, 1892.

[7] Emmons, S. F., op. cit., p. 324.

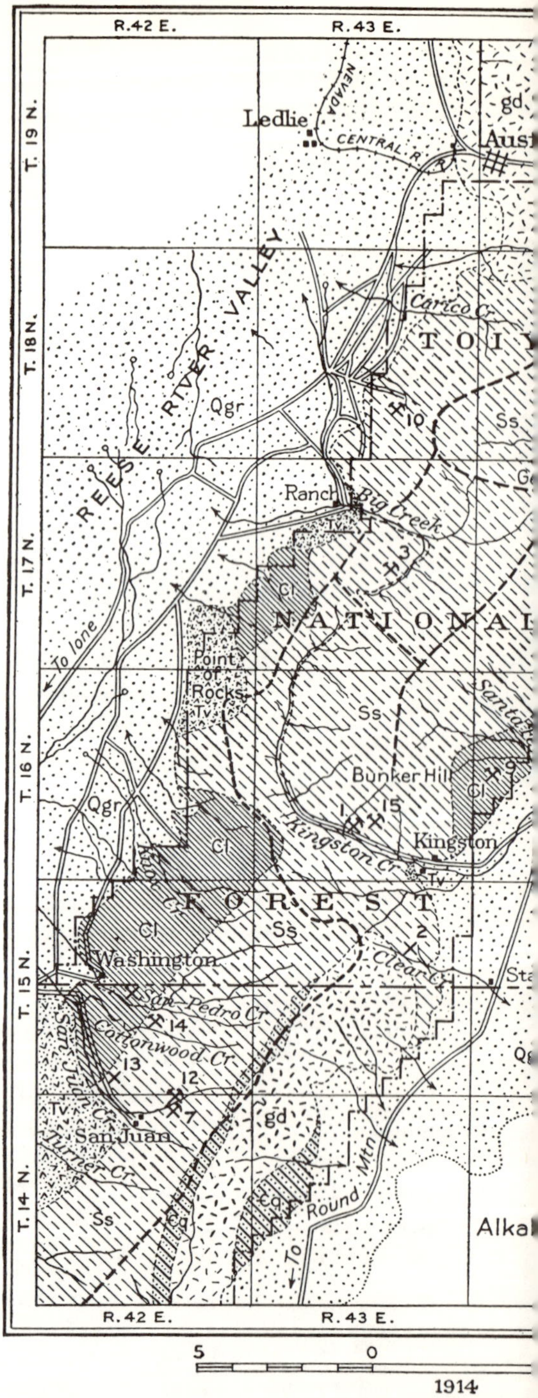

GEOLOGIC SKETCH MAP OF A PORTION O
OF AUSTIN, LANDER CC

The geology is shown essentially as mapped

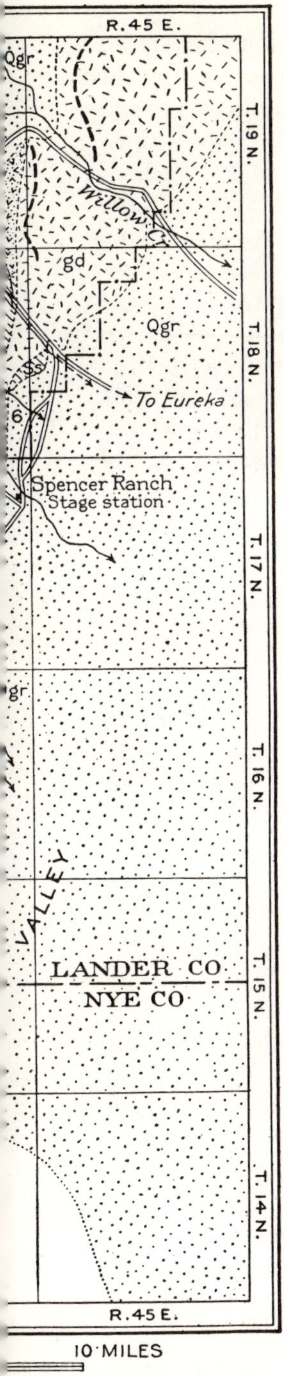

LEGEND

SEDIMENTARY ROCKS

QUATERNARY
- Qgr — Gravel

CARBONIFEROUS
- Cl — Dark-gray limestone

SILURIAN ?
- Ss — Slate and shale, and some limestone

CAMBRIAN ?
- €q — Quartzite and some thin limestones

IGNEOUS ROCKS

TERTIARY ?
- Tv — Volcanic rocks, largely rhyolite

CRETACEOUS OR TERTIARY ?
- gd — Granodiorite (*Intrusive into Paleozoic sediments*)

- Divides
- Mine
- Prospect

LIST OF MINES AND PROSPECTS

1. Bi-Metallic
2. Bowman
3. Bray
4. Bull Dog Jack (Moss)
5. Deer Point
6. Galliger group
7. Henry George (Morning Star)
8. Limelight
9. Mother Lode
10. Pine
11. Smoky Valley
12. St. Louis (Richmond)
13. Tiger
14. Werner (Washington)
15. Victorine

YABE RANGE SOUTH
EV.

eth Parallel Survey.

conformably and without any marked line of division. Emmons further noted that this shale series was quite similar to if not the same as that exposed at Belmont, Nev., in which Gilbert[1] found Silurian fossils. In the report of the Fortieth Parallel Survey these slates were referred to the Carboniferous on account of the lack of fossil evidence, but it would seem better to provisionally refer them to the Silurian on account of their lithologic similarity to the known Silurian slates at Belmont.

Limestone.—Three areas of compact dark blue-gray to light-gray crystalline limestone are exposed on the flanks of the range, as shown in Plate XIII (p. 114). The largest area of somewhat crystalline but otherwise unmetamorphosed limestone is on the west side near Washington; the narrow strip along the east side of the range between Tar Creek and Kingston is much metamorphosed and cut by numerous stringers of calcite. The limestones rest conformably upon the shale. Emmons[2] found *Fusulina cylindrica* and Syringopora, fossils of Carboniferous age, in these limestones at Santa Fe Canyon.

IGNEOUS ROCKS.

INTRUSIVE ROCKS.

Granodiorite.—There are three rather large areas shown on Plate XIII which are underlain by a light-gray siliceous igneous rock, the most common phase of which is granular, though in some places, as at Birch Creek, a porphyritic texture is noted. The typical rock is composed of microcline, orthoclase, quartz, oligoclase, greenish biotite, colorless muscovite, and magnetite, named in the order of their abundance. Hornblende is locally present as an accessory but is not typical of this intrusive. The granular rock mineralogically resembles the granodiorite of the Sierra Nevada, with which it is probably to be correlated.

At the mouth of Birch Creek the eastern border of the "granite" mass is nearly white and has a coarse porphyritic texture. Large well-developed orthoclase crystals, the largest 1 inch in diameter, and rounded quartz grains are the phenocrystic minerals, set in a coarse granular matrix of microcline, orthoclase, oligoclase, quartz, muscovite, and biotite. In some parts of this porphyritic granodiorite mass there are small phenocrysts of biotite as well as quartz and orthoclase. This rock is cut by veinlets of quartz, some of which carry metallic minerals, and the surface of the whole hill is more or less iron stained.

The granodiorite is intrusive into the Paleozoic sediments, as is clearly seen at the mouth of Birch Creek, where the limy shales and

[1] Gilbert, G. K., Geology: U. S. Geog. Surveys W. 100th Mer., vol. 3, p. 180, 1875.
[2] Emmons, S. F., op. cit., pp. 323, 335.

intercalated limestones have suffered contact metamorphism that has produced zones consisting of quartz, diopside, colorless garnet (grossular), some pinkish iron-bearing garnets, and a small amount of epidote.

Granular dikes.—Two varieties of dikes, too small to be shown on a map of the scale of Plate XIII, cut both the "granite" and the sedimentary rocks. On the east side of the range these dikes usually strike about parallel to the trend of the mountains and dip at different angles. Many of those in the sedimentary rocks follow the bedding in a rough manner, though several nearly vertical dikes were noted. They are narrow, few of them exceeding 50 feet in thickness, and most of them are less than 25 feet wide. On the western flank of the mountains there do not seem to be as many dikes cutting the sediments, though in Big Creek one siliceous dike was seen.

A black fine-grained granular rock is most widely distributed as dikes. Some of these dark dikes are impregnated with pyrite, which in some of the dikes is unaltered, but in the majority of them is altered to hematite or limonite. This dike rock is composed of calcic feldspar, probably labradorite, with light-greenish augite, and locally some hypersthene; the augite in one thin section showed slight alteration to green hornblende. Brown biotite, though not abundant, is present in this rock, and locally there is a small amount of olivine. Magnetite is a rather abundant accessory mineral. In the thin sections of this type of rock the small interstitial spaces between the main constituents are filled with an intergrowth of quartz and orthoclase. This rock, to judge from the thin section, ranges between a quartz diorite and a quartz-olivine gabbro.

Glassy dikes.—The dikes of the second type are light-gray, much altered, apparently somewhat glassy rocks, heavily impregnated with pyrite. They are not widely distributed and seem to be confined to the region between Big Creek and Tar Creek. In the field they were called rhyolites, a classification which can not be verified on account of the altered condition of the specimens in hand. Two thin sections show this rock to be completely altered to an aggregate of small crystals of sericite, quartz, and calcite.

It was the opinion of the writer while in the field that both of the dike rocks were differentiation products of the granodiorite intrusion, and for the dark dikes this theory seems to be substantiated by microscopic study. The altered light-colored dikes are not so clearly of this origin. The structure of the rock in thin section seems rather to point to its being related to the rhyolite flows.

EXTRUSIVE ROCKS.

There are five areas of extrusive igneous rocks within the area shown on Plate XIII (p. 114). The largest of these areas west of Austin is underlain by a brownish-purple, somewhat vesicular rock, carrying phenocrysts of smoky quartz, small orthoclase crystals, and muscovite in a microcrystalline feldspathic base. At the summit of Lander Hill, which is called Mount Prometheus, this rhyolite, where it is in contact with the "granite," passes into a black obsidian.

The next largest area lies south and west of San Juan Creek, southwest of Washington. The limestones are here capped by cream to buff colored rhyolites and rhyolitic tuffs, which have a gentle west-southwest dip.

Point of Rocks, the third area in the left center of Plate XIII, was not visited by the writer but is described by Emmons[1] as—

composed of propylite; * * * the limestones and slates are found to be dipping away from it on either side and somewhat altered at the contact. The main central mass is a hornblendic propylite [andesite]; * * * along the summit of the ridge this rock has a distinctly columnar structure * * *. It is traversed here by a dike of compact dark rock, resembling andesite but carrying large rounded grains of limpid quartz * * *. This central mass of hornblendic propylite is inclosed on the west and south by a body of quartz propylite * * *. This quartz propylite is a later phase of the eruption, being separated from the main body on its southern edge by a tufa-like breccia.

At the mouth of Big Creek, north of Point of Rocks, there is a small area of reddish-purple rhyolite overlying the slates and limestones and forming the low hills for a short distance south of the canyon. At the base of this rhyolite there is a zone composed of fragments of limestone and slate in a matrix of the rhyolite.

Just south of Kingston Canyon, on the east side of the range, there is a small exposure of reddish rhyolite, in some places resembling a breccia. Other exposures look like slates but have distinct quartz and feldspar phenocrysts in a base showing flow structure.

These extrusive rocks are probably to be correlated with the Tertiary volcanic rocks so common in Nevada, though no definite information as to their age was obtained in the field.

STRUCTURE.

The general structure of the range is anticlinal, as was noted by Emmons,[2] who says the elevation is—

the result of the action of forces of contraction or compression acting in two directions, laterally, or at right angles to the general line of elevation, and longitudinally, or parallel to that line. The lateral or main forces of upheaval have produced anticlinal and synclinal folds, whose axes would have the general direction of the range. The effect of the longitudinal compressions has been a disturbance of these conditions, causing a deviation from the general direction and a crumpling and general dislocation of the strata * * *.

[1] Emmons, S. F., op. cit., pp. 338–339.
[2] Idem, pp. 325–326.

The central mass of sedimentary rocks comprising the double-ridge system of Globe, Bunker Hill, and Big Creek peaks * * * forms a general anticlinal fold; * * * here the effect of the longitudinal compression is most distinctly seen, since the axis of the fold has an extreme variation in direction from about N. 35° E. on the northern end to N. 20° W. on the southern.

The sections in figure 4 are taken from Plate XXVI of Emmons's report [1] and give an idea of the broader features of the structure. Emmons describes these sections as follows:

A, on a line drawn through Mount Prometheus E. 13° S., shows the relation of the granite and rhyolite bodies at that point.

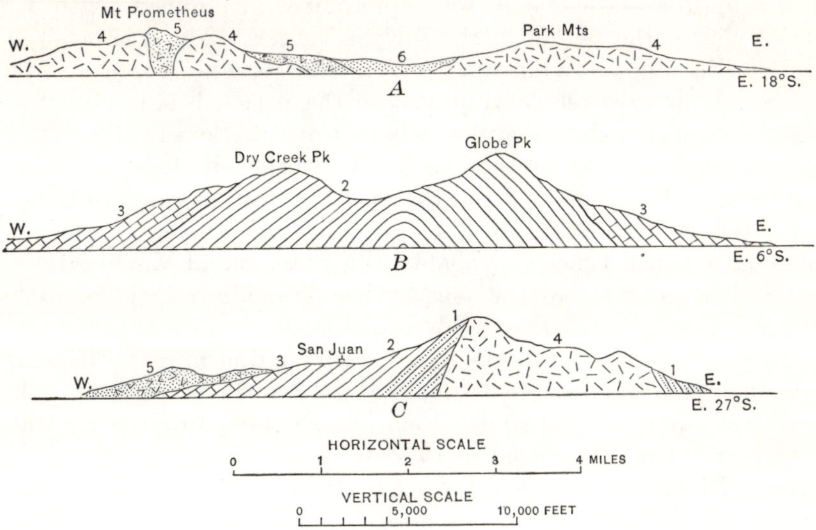

FIGURE 4.—Structure sections of the Toyabe Range (after S. F. Emmons, Mining industry: U. S. Geol. Expl. 40th Par., vol. 3, Pl. XXVI, 1870). 1, Cambrian (?) quartzite; 2, Silurian (?) slate and shale; 3, Carboniferous limestone; 4, granodiorite; 5, volcanic rocks; 6, gravels.

It seems questionable whether the rhyolite of Mount Prometheus is in the form of a dike as shown. There are undoubtedly small rhyolite dikes that cut the granite near the summit of this mountain, but the main mass of the peak seems to be a flow rather than a dike as shown.

B, on a line drawn through Globe and Big Creek peaks, E. 6° S., shows the main anticlinal fold, though it can not serve for an accurate representation of the thickness of the strata, since their strikes vary so much that it is impossible to choose a line which shall cut them at right angles.

This section is very much generalized and locally the longitudinal compression has been great, so that the beds are much contorted and have a strike closer to east and west than would appear from the section.

[1] Emmons, S. F., op. cit., pp. 327, 328.

C, on a line drawn through San Juan E. 27° S., shows a section of the central granite body, the slates and quartzite resting upon it, and the rhyolite on their flanks.

ORE DEPOSITS.

HISTORY AND PRESENT CONDITIONS OF MINING.

Following the discovery of the veins on Lander Hill in 1862 the prospectors searched practically every canyon and ridge of the Toyabe for the precious metals.

Smoky Valley, on the eastern side of the mountains, was worked in the middle sixties,[1] but was apparently abandoned when Emmons[2] visited the region in the early seventies. The Victorine and Bi-Metallic mines in Kingston Canyon were located in 1862 and 1863 and saw their greatest development in the period between 1872 and 1887, since which time they have been worked sporadically without marked success. The Mother Lode was worked in 1869 and a mill was built at Kingston to treat the ores.[3] Other prospects scattered along the western edge of Smoky Valley have been located and abandoned many times. Probably the most recent discoveries are the gold-silver veins in the porphyritic granodiorite at the mouth of Birch Creek, which were made in 1910.

In 1875 Raymond[4] reports four amalgamation mills in Kingston Canyon, none of which had been operated for several years. The largest of these, the Stirling, which has 20 stamps driven by water power, was being remodeled in 1912 by lessees.

Reese River valley was followed southward by the earliest prospectors. The mines at San Juan and Washington were located in 1862 and 1863, but the veins are small and their ore is base, so that they have not received much attention. In the vicinity of Big Creek there are a number of antimony prospects, one of which, the Bray, was located in 1864 as a silver mine. It was not until 1891 that the stibnite was recognized. During the seven years following this discovery development exposed a considerable quantity of antimony ore, which was shipped by the owners and leasing companies. The Pine mine, north of Big Creek, was discovered in 1890 and a small quantity of antimonial ore has been shipped from it.

Very little work was under way in this part of the range in the summer of 1912. At the Moss mine a small force was working, and the Kingston Canyon properties were being put in shape for development by a leasing company, which was later reported to have sus-

[1] Browne, J. R., Mineral resources of the States and Territories west of the Rocky Mountains, 1867, p. 413.

[2] Emmons, S. F., Mining industry: U. S. Geol. Expl. 40th Par., vol. 3, p. 333, 1870.

[3] Raymond, R. W., Mineral resources of the States and Territories west of the Rocky Mountains, 1868, p. 83.

[4] Raymond, R. W., Statistics of mines and mining in the States and Territories west of the Rocky Mountains, 1875, p. 329.

pended operations. On a number of scattered claims the owners were doing the annual assessment work, but as a whole the region showed little activity.

CHARACTER OF THE DEPOSITS.

The ores are very similar throughout this range. Pyrite is common but as far as known is not abundant and carries very small quantities of valuable metals. The principal constituents of value are galena, usually containing some antimony; tetrahedrite or "gray copper," which is invariably silver bearing in this district; dark iron-bearing sphalerite, and locally small quantities of chalcopyrite. The stibnite veins in the vicinity of Big Creek carry a little argentiferous gray copper but are quite free from the lead and zinc materials. At one place on the east side of the range quartz veins in granodiorite carry molybdenite, chalcopyrite, and sphalerite. The surface ores of the silver veins consist of iron-stained quartz that contains different quantities of blue and green copper carbonates and silicate, depending on the content of gray copper in the original ore. Argentite (silver sulphide) was found in a few of the mines and was probably of some importance in the old stopes of the Victorine vein in Kingston Canyon. The gangue mineral of all these veins is vitreous quartz, generally white, though in some places gray.

The veins were deposited in open fissures, and crustification is common in all of them. Some of the deposits are best described as lodes, as they consist of a network of small interlacing quartz veinlets cutting the formation in all directions. This lode structure is most commonly seen in the contorted shales. True veins, however, are found in this formation as well as in the "granite." The true veins are not large, in few places exceeding 4 feet in thickness and generally ranging from a few inches to 2 feet in width. The lodes, on the other hand, have a width of from 25 to 250 feet.

The Big Smoky deposit and some small copper deposits northeast of the Big Smoky mine have certain characteristics in common with contact-metamorphic deposits, though in all places they have more the form of veins. At the Big Smoky the crustified quartz vein occurs in a belt of limestone altered to a quartz-diopside-garnet rock near the contact of intrusive granodiorite porphyry.

MINES ON THE WEST SIDE OF THE RANGE.

VICINITY OF BIG CREEK.

Pine mine.—The Pine mine (No. 10, Pl. XIII, p. 114) is on the crest of a west spur of the range, about 10 miles by road south-southwest of Austin. The ledge, reported to have been discovered in 1890, is developed by an inclined shaft said to be over 230 feet deep, a crosscut tunnel, and numerous open cuts and pits. The ledge strikes N. 55° W. and dips 50° SW., being about 40 feet wide and traceable along the strike for nearly 400 feet. The ledge stands from 15 to 20 feet above the surrounding sedimentary rocks, which strike northwest and dip southwest at medium angles. These sedimentary rocks are dark siliceous shales and slates containing some thin quartzite beds and a few thin beds of dark-gray limestones.

The ledge appears to be a silicified fault breccia cemented by white vein quartz intergrown with stibnite. The mineralization seems to have taken place along nearly vertical fractures that strike north, and cut the breccia nearly parallel to the fault at the southeast, which has cut off the ore body. Slickensided faces parallel to the dip of the ledge occur in it in many places, but are particularly strong along the southwest hanging-wall side.

To judge from the dump, the deeper workings of the incline, which was badly caved at the surface, are in limestone, and the mineralization does not appear to have been so strong as at the surface.

The ore is largely white quartz and stibnite, the latter mineral usually occurring in small crystalline masses intergrown with the quartz. Some masses of pure stibnite were seen as large as a foot in diameter. On the surface there is a soft yellowish coating of antimony oxide on some of the ore, though this is not common. In some of the open northward-striking fractures cellular grayish quartz occurs, coated with a fine black powder of manganese oxide.

Bray mines.—The Bray claims (No. 3, Pl. XIII, p. 114) are on the west side of the south fork of Big Creek about 5 miles from the mouth of the canyon and 1½ miles south of the Kingston Canyon divide. These claims were located for silver about 1864, but were abandoned and relocated in 1891 by Joseph Bray. They were worked at intervals up to 1898, and considerable sorted ore, carrying 50 per cent antimony, is said to have been shipped to Swansea, Wales, by way of San Francisco.

There are two mines, the Bray and the Pradier, located respectively on the crest and north flank of a zone of folding and faulting that strikes N. 20° W. The thin-bedded dark siliceous slates and reddish shales of the Silurian(?) are folded into little anticlines and synclines parallel to the axis of the main fold, and the beds have been faulted along most of the folds. On the crest of the main anticline at the Bray mine the contorted shales and slates are cut in all directions by white quartz stringers from a fraction of an inch to 8 inches in width. Both massive and crystalline stibnite is intergrown with the quartz. Few of the crystals are large, but the massive mineral commonly occurs in bunches, the largest of which are 8 inches in diameter. The stringers of ore have been faulted in some places as much as 8 inches by postmineral movement, though they seem to have been deposited after the main crushing and folding along this zone. The main development on the Bray claim is a tunnel and numerous drifts, said to aggregate 900 feet of work. It was under water and could not be entered. On the dump of this tunnel there is some much-altered pyritized porphyry that seems to be an altered rhyolite.

The Pradier claim is situated on the north side of the folded zone, about 1,000 feet from the Bray. It is developed by several short tunnels, only two of which could be entered. These tunnels are said to have been the work of the last lessees, and are so small that one has to crawl on hands and knees in most places. The ore here occurs as thin tabular, very irregular lenses of quartz and stibnite that lie parallel to the bedding of the shales, which strike approximately N. 20° E. and dip north at low angles. Associated with the stibnite there is a little gray copper, which has been altered to the green and blue carbonates, and the silicates on the surface.

In the canyon bottom just east of the Bray mines there is a small brick furnace, which was erected a few years ago for treating ores. It was apparently a failure.

VICINITY OF WASHINGTON.

San Pedro Canyon.—There are several prospects about one-half mile east of Washington in San Pedro Canyon. These prospects are in limestones that strike N. 40°–60° W. and dip southwest at steep angles and are cut by small rather indefinite dikes of granodiorite. The prospects contain quartz veins from 2 to 4 feet in width that are about parallel to the structure of the inclosing rock. The visible metallic minerals are pyrite, galena, and small quantities of gray copper which has altered and given copper carbonate and silicate stains to some of the croppings.

Werner mine.—The Werner mine (No. 14, Pl. XIII, p. 114), known as the Washington property, is situated on the ridge south of San Pedro Canyon, about a mile southeast of Washington. The deposit is located just east of the contact between the limestone and shale in beds consisting largely of lime-shale and quartzite. North of the ore body there is a small, roughly elliptical mass of granodiorite exposed in a crosscut tunnel. The sedimentary rocks strike a little north of east and dip to the north at low angles, showing a slight displacement transverse to the general strike and dip in this vicinity. Emmons[1] says of this deposit:

> In these limestones between Washington (San Pedro) and Cottonwood canyons is the New Hope vein, which when visited showed only the croppings of a large body of quartz, apparently striking north and south, carrying ores of

[1] Emmons, S. F., Mining industry: U. S. Geol. Expl. 40th Par., vol. 3, p. 339, 1870.

silver, combined with the yellow oxides of lead and blue carbonate of copper. This was said to be very rich, yielding $800 to the ton.

On the surface the croppings do seem to strike north and south, yet the upper development work shows that the quartz is associated with an eastward-striking fracture that dips 10° S. This siliceous body is a quartzite breccia recemented by quartz, and stained with iron on the surface. Near the fracture there is a small amount of gray copper and galena, which are largely altered to copper carbonates and silicates and the yellow lead carbonate. There is also a small quantity of a soft black mineral that seems to be silver glance (argentite). At a depth of 30 feet below the surface the hard quartz gives way to a slightly consolidated breccia consisting of fragments of shale, quartzite, limestone, and a little vein quartz. This zone of brecciation is 10 feet wide, has an easterly strike, and dips 10° S. As far as known only one shipment has been made from this property. This shipment was tested and is said to have carried about $10 a ton in silver and lead with a little copper.

There are several lead and silver prospects in the shale area northeast of Washington. The white quartz veins, said to be quite small, have a general north-northeast trend.

SAN JUAN CANYON.

Tiger group.—The Tiger group (No. 13, Pl. XIII, p. 114) lies on the north side of San Juan Canyon, about 3 miles above the Reese River valley, just east of the contact of the shale and limestone. The development, consisting of open cuts, shafts, and a 300-foot crosscut tunnel, exposes a fracture cutting dark siliceous shales that strike N. 55°-60° W. and dips 70° N. The sedimentary rocks are disturbed for 15 feet south of the fracture and are cut by small stringers of quartz and pyrite, which are also disseminated through the crushed rock. With the pyrite occur very small quantities of galena and dark sphalerite. The croppings are stained yellow by the oxidation of pyrite, giving free sulphur. The last 60 feet of the tunnel penetrate the overlying blue-gray crystalline limestone.

Henry George vein.—The Henry George vein (No. 7, Pl. XIII, p. 114) crosses the east fork of San Juan Canyon about 1 mile east of the forks. This vein strikes north and dips 50° E., cutting across the westward dipping dark shales. It ranges from 4 to 6 feet in width and consists of white vitreous quartz, in some places frozen to the walls and in others separated from them by thin clay partings. There are numerous little quartz stringers making into the walls both parallel to the stratification of the sediments and cutting across it. The vein is opened in several places for about 2,000 feet along its strike, but most of the work is inaccessible. In a short tunnel and winze on the south side of the creek the vein is well exposed. The ore minerals are antimonial galena, said to be silver bearing, intergrown with dark iron-bearing sphalerite and a very small quantity of pyrite.

St. Louis and Richmond.—The St. Louis and Richmond workings (No. 12, Pl. XIII, p. 114) are at a barometric elevation of 8,575 feet, in the east fork of San Juan Canyon, on a vein that strikes N. 15° E. and dips 70° W., parallel to the bedding of the sedimentary rocks. The vein ranges from 4 to 14 inches in width, and lies in a light-gray bed of limestone interstratified with the dark Silurian(?) shales. The ore on the dump shows the vein to be well banded, sulphides being deposited next the walls, followed by bands of calcite and siderite, and finally the central band of drusy white quartz. The sulphides associated with the quartz and siderite are pyrite, dark sphalerite, galena, and arsenopyrite, all apparently being deposited contemporaneously, though most of the ore shows that pyrite is a little older, being deposited in quite pure streaks next the walls. The zinc content is variable, but is excessive in most of the material on the dump. As far as could be learned only one small shipment of ore was made from these properties, and that was for testing. It is said to have averaged $34.50 in lead and silver with very small quantities of gold.

MINES ON THE EAST SIDE OF THE RANGE.

VICINITY OF BIRCH CREEK.

Smoky Valley mine.—The Smoky Valley mine (No. 11, Pl. XIII, p. 114) is on the ridge south of the mouth of Birch Creek, at a barometric elevation of 7,000 feet. The development consists of several shafts and a crosscut tunnel, all of which have been abandoned for a number of years, to judge from appearances.

Emmons[1] says:

> This tunnel has been cut over 800 feet into the granite spur * * * for the purpose of reaching the Big Smoky vein. * * * The workings on this vein disclose a large irregular body of quartz and calcspar from 15 to 20 feet thick, stained by graphite; it would seem to be at the contact of the slates and granite, since these rocks are cut in alternate bands by the tunnel.

The main ore body consists of white quartz and a little calcite. It strikes N. 65° E. and dips northwest at steep angles, lying about 200 feet southeast of and parallel to the contact of the granodiorite and the shale, though small dikes of granodiorite occur near the vein. Near some of the dikes the calcareous shales are altered to masses of diopside, pink garnet, and quartz. The ore at the surface

[1] Emmons, S. F., op. cit., p. 333.

is only slightly iron stained, but some quartz on the lower dumps carries fairly abundant disseminated arsenical pyrite.

Deer Point group.—The Deer Point claims (No. 5, Pl. XIII, p. 114), 10 in number, are situated on the ridge south of Birch Creek and west of the Smoky Valley mine at elevations ranging from 7,300 to 7,900 feet. The country rock is coarse porphyritic granodiorite, containing very small amounts of ferromagnesian minerals. This rock is cut by two series of joints. One set, which stands nearly vertical, strikes N. 65° E.; the other strikes N. 20° W. and dips to the west at low angles. Along some of these joints there has been a little mineralization, which has resulted in the formation of narrow quartz stringers carrying pyrite, galena, and sphalerite. In some of these stringers only pyrite has formed, whereas in others galena and sphalerite are the principal metallic minerals. The granodiorite near the mineralized joints contains a small amount of disseminated pyrite. At one place a zone about 20 feet wide is cut by a network of small vertical joints and is deeply iron stained and sericitized.

Other prospects.—In the granodiorite 3 to 4½ miles northwest of Spencer ranch (see Pl. XIII, p. 114) there are a number of prospects on small irregular quartz veins, carrying pyrite, gray copper, galena, and sphalerite. The veins are in most places frozen to the walls, and in only a few places have they been opened by workings larger than cuts and pits. On the surface the veins are usually somewhat iron stained and locally show the brilliant blue and green copper carbonate and silicate stains. These veins in general strike about east and west and dip north at medium angles, though some strike N. 15° W. and stand vertical.

About 4½ miles north-northwest of Spencer ranch and a mile west of the stage road there are some molybdenum prospects in granodiorite. The principal vein so far exposed strikes N. 50° W. and dips 30° S. It is developed by an incline about 50 feet deep. It ranges in width from a few inches to 18 inches and is frozen to the walls. The gangue, a vitreous white quartz, constitutes about 80 per cent of the vein. In this gangue there are scattered small flakes of molybdenite the largest three-fourths of an inch in dimension, as a rule associated with a cream-white micaceous substance and a small quantity of chalcopyrite which is largely altered to limonite and copper carbonates.

In the low foothills about 2½ miles north of Spencer ranch that are composed of dark shales with some impure limestone beds, there are a number of copper prospects (No. 6, Pl. XIII, p. 114), largely owned by Mrs. Galliger and John Cahill, of Austin. These prospects are not far from the granodiorite contact, and the sedimentary beds stand nearly vertical and have north-northeast strike. The limestone beds have been metamorphosed into a hard quartz-garnet-epidote rock. In a few places this altered limestone is impregnated with dark sphalerite and chalcopyrite. The sulphides are found at the surface, but oxidation has extended to at least 70 feet, the greatest depth yet reached in any of the workings. The ore occurs in the metamorphosed limestone in small irregular lenses, few of which are over 4 feet wide and are generally between 12 and 14 inches wide.

VICINITY OF TAR CREEK.

Bull Dog Jack group.—The Bull Dog Jack group of eight claims (No. 4, Pl. XIII, p. 114), located on the divide between Tar Creek and the north fork of Big Creek, is best reached by a steep pack trail from the Big Creek side. This group is also known as the Moss mine, from the name of the locator.

The metamorphosed black shale series is here cut by two sets of irregular white quartz veins carrying small amounts of argentiferous galena, dark sphalerite, and locally a little pyrite. The veins of one set, striking N. 85° E. and dipping 75°–80° S., are generally frozen to the walls, whereas those of the other set, which strike N. 15° W. and dip steeply to the east, are as a rule separated from the walls by thin gouge. There is a 150-foot drift on this ground that shows both series of veins and a shaft on an eastward striking vein at the summit of the hill. There is apparently only a small amount of available ore in these veins and as yet none has been shipped, though it is said that picked samples run as high as $100 a ton, $2 to $4 of which is in gold and the remainder in silver and lead.

Cahill claims.—A few hundred feet east of the Bull Dog Jack, John Cahill, of Austin, has a claim called the Limelight, on which the quartz lode shows essentially the same mineralization as that on the group just described. The sedimentary beds in this vicinity strike N. 45° W. and dip 70° SW. The lode consists of a number of quartz stringers 4 inches to 2 feet in width, cutting a calcareous shale that has a general strike of N. 60° W. and a dip of 40° S. Postmineral movement has crushed the lode to some extent, and white mica with quartz has been deposited in the interstices of the breccia.

On the north side of Tar Creek, about halfway down from the summit, Cahill owns some claims on quartz pyrite stringers in which the values are said to be largely in gold. They occur in the shale series near a basic dike.

Other prospects.—A small amount of development work has been done near the mouth of the canyon next north of Tar Creek (see Pl. XIII) on a 25-foot basic dike that is nearly parallel to the stratification of the sedimentary rocks, which here strike northeast and dip southwest at medium angles. A small amount of pyrite is disseminated in this dike, and at the surface the contact zone is lightly stained by iron and copper carbonates.

VICINITY OF KINGSTON CANYON.

Victorine mine.—The Victorine mine (No. 15, Pl. XIII, p. 114), one of a group of 11 claims, 6 of which are patented, has had a checkered career since its discovery in 1862. It has been opened and shut down countless times by various lessees, none of whom has ever been able to make a success of the property. The ore occurs in pockets of quartz irregularly scattered in a bed of silicified limestone through a vertical width of 25 feet. The limestone is interbedded with black siliceous shales and contorted slates. This zone is parallel to the sediments, trending east and west and dipping from 5°–35° N. It is traceable along the north side of Kingston Canyon for about 2 miles and has been opened by a number of long, inclined tunnels. As noted by Emmons,[1] there is a 50-foot basic dike about 200 feet below the vein that is also parallel to the structure of the sedimentary rocks.

The ore at the surface consists of rather porous, iron-stained quartz carrying limonite and copper carbonates. With depth the quartz is both white and dark gray, and has pyrite, galena, sphalerite, and tetrahedrite irregularly scattered through it. Argentite and silver chloride are said to have been found in the old Victorine tunnels, which could not be entered.

Several mills have been built to treat this ore. The largest, at the mouth of Kingston Canyon, was being remodeled in September, 1912, to treat the ore by cyanide methods.

There are no exact figures of production for this mine, though it is said that a considerable tonnage was hauled to Austin for treatment before the mills at Kingston were built.

Bi-Metallic vein.—The Bi-Metallic vein (No. 1, Pl. XIII, p. 114) is a westward continuation of the Victorine and is similar to it in all respects.

Mother Lode.—The Mother Lode (No. 9, Pl. XIII, p. 114) is on the point of the ridge north of Kingston, about 1,500 feet above Smoky Valley. The large white quartz croppings are clearly seen from the stage road. The ore is similar to that of the Victorine, except that it seems to have more pyrite. The mine was not working in 1912, and it is said that it never was a financial success on account of the low grade of the ore.

Bowman group.—The Bowman group of 18 locations (No. 2, Pl. XIII, p. 114) is on the north side of Clear Creek Canyon, the first large break in the mountains south of Kingston. The country rock in this region is the black slate intercalated with dark limestones and some light-colored quartzitic beds. These sediments are much crumpled into schistose rocks along a zone of crushing that strikes

[1] Emmons, S. F., op. cit., p. 336.

north to N. 25° E. They are cut by at least two northward-striking basic dikes in the vicinity of the claims.

The main development work is a 350-foot crosscut tunnel that runs N. 55° E. into the hill toward the croppings of the largest of the veins seen on the surface. In this crosscut the crumpled lime schists are cut by a multitude of small irregular white quartz stringers running in all directions, though the larger veins have a general northerly trend and low angles of dip, either east or west. The quartz stringers carry a small amount of iron oxide that seems to have altered from pyrite. In the last 250 feet of the crosscut the quartz stringers are particularly abundant and the schist between them has been mineralized, carrying small crystals of pyrite that are rather abundant. This whole body is said to assay 4 ounces of silver and 60 cents in gold to the ton.

GRANITE DISTRICT, MINERAL COUNTY, NEV.

LOCATION AND PRESENT ACTIVITIES.

The Granite district (No. 12, Pl. I, p. 18) is situated along the summit of the Wassuk Range about 6 miles west-northwest of Schurz in an air line. The region is shown in the northwest corner of the Hawthorne topographic sheet of the United States Geological Survey. Schurz, the supply point on the Hazen-Tonopah branch of the Southern Pacific Railroad, is about 15 miles by good roads from the mines. The district was visited in 1908 by F. L. Ransome,[1] who published a brief description of it.

There were formerly two small settlements in the district—Mountain View, on the west of the summit, and Granite, near the head of an eastward-draining canyon. In the summer of 1912 both places were abandoned. At the Beach mine, belonging to the Yerington Mountain Copper Co., a few men were working in July, 1912, and it is said that a little work was being done at some copper claims about 3 miles south of the Beach mine. The only recorded production from the district was made in 1909, when about $4,500 in gold and silver were recovered.

There are two wells in the canyon bottom near Granite, but the water is not particularly good and the supply is small. There are also two small depressions southwest of Black Mountain, a basaltic peak south of the Beach mine in which some water is held for a time after the spring thaw. The drinking water used at the Beach camp is hauled from Schurz.

[1] The Yerington copper district, Nev.: U. S. Geol. Survey Bull. 380, pp. 118–119, 1909.

TOPOGRAPHY.

The Wassuk, or, as it is locally known, the Walker River Range, has a very steep eastern escarpment overlooking Walker Lake, along which Russell[1] has postulated a pre-Quaternary fault. The mountains attain a height of 8,000 feet in a distance of 2 miles west of the edge of Walker Lake, whose elevation is 4,050 feet above sea level. The fault scarp is modified by erosion, which has resulted in the cutting of many short, deep canyons that have very steep gradients from the summit of the range to Walker Lake. Practically all these canyons are dry except for short periods after the spring thaw. The western side of the range, on the contrary, has a much gentler slope and in general a smoother topography than the eastern side. At the south end of Mason Valley there are groups of low hills between the flats and the range which tend to make the slope appear less abrupt.

GEOLOGY.

Granodiorite.—The main mass of the Wassuk Range is composed of a gray granular igneous rock composed of feldspar, quartz, biotite, and either augite or hornblende with accessory magnetite, apatite, and zircon. Ransome refers to it as "granodiorite or quartz monzonite." As the plagioclase appears to be a little more abundant than the orthoclase the rock is here called granodiorite, though it may be nearer a quartz monzonite in chemical composition.

The granodiorite is cut by east-west dikes of granodiorite porphyry, aplite, and augite andesite. The granodiorite porphyry has essentially the same mineral composition as the granodiorite. Phenocrysts of plagioclase and orthoclase, the largest being one-eighth inch in diameter, are set in a granular groundmass of the two feldspars with quartz, biotite, augite, and hornblende. These dikes are not abundant and are usually from 10 to 20 feet wide. The aplite dikes are quite narrow, few of them exceeding 5 feet and most of them being 2 feet or less in width. The wider dikes are pinkish and are composed of small granules of quartz and orthoclase. Some of the narrow dikes are composed of glass with more or less crystalline quartz and feldspar, and the edges of some of the wide dikes show the same glassy base. A few small basic dikes having an eastward strike cut the granodiorite on the summit between the Beach and Big Twenty mines. This dark porphyry has a groundmass composed of andesine laths, augite, now largely chloritized, and magnetite, in which there are small andesine and augite phenocrysts. The rock is quite similar in mineralogic composition to the lowest

[1] Russell, I. C., Geological history of Lake Lahontan: U. S. Geol. Survey Mon. 11, p. 27, map opp. pp. 28 and 70, 1885.

flow rock on Black Mountain. A thin section of the later rock is much fresher, showing the augite to be very light colored, and that the rock contains abundant accessory apatite.

Where the road from Schurz to Granite enters the mountains it skirts the southern side of White Mountain, a mass composed of granodiorite so cut by white siliceous dikes that the hill is almost white in color. These dikes appear to be largely quartz and orthoclase, which Spurr[1] has called alaskite-aplite.

This large mass of granodiorite is probably to be correlated with the granodiorite intrusions of the Sierra Nevada, which occurred in late Cretaceous or early Tertiary time.

Andesites and rhyolites.—On the western side of the mountains flow rocks obscure the granodiorite.[2] These flow rocks were seen by the writer only in the vicinity of Mountain View, where gray and reddish-colored andesites and rhyolites occur and, according to Ransome, are capped by basalt south of Mountain View. These flows are probably of Tertiary age.

ORE DEPOSITS.

Occurrence.—The veins of the Granite district occur in the granodiorite in some places near the intrusive dikes, though they do not seem to be closely related to them. There are two veins near Granite which have been developed to some extent, but only one of these veins was being worked in July, 1912. Ransome[3] gives some notes on these deposits to which but little can be added.

Beach vein.—The Beach vein, controlled by the Yerington Mountain Copper Co., is opened by a drift tunnel whose mouth is 1½ miles south of Granite, near the summit of the range. In July, 1912, the tunnel was 650 feet long and the air connection 143 feet deep. A small amount of stoping on the vein within 300 feet of the portal and on an ore shoot 190 feet from the face of the drift comprise the workings. The country rock is granodiorite that is somewhat sericitized near the vein. A dike of aplite cuts the granodiorite at the mouth of the tunnel and 300 feet from the mouth a 10-foot northward-striking dike of aplite is cut by the vein.

At a point 140 feet from the face of the tunnel there is a small body of alaskite that seems to be intrusive into the granodiorite on the hanging wall of the vein. The vein is covered by the augite andesite flows about 1,500 feet southwest of the tunnel mouth.

---

[1] Spurr, J. E., Descriptive geology of Nevada south of the fortieth parallel and adjacent portions of California: U. S. Geol. Survey Bull. 208, p. 115, 1903.

[2] Idem, p. 116.

[3] Ransome, F. L., The Yerington copper district, Nev.: U. S. Geol. Survey Bull. 380, pp. 118–119, 1909.

The croppings of the vein, from 8 to 25 feet wide, stand above the surface, striking about N. 50° E. Underground the average strike of the vein is N. 35° E. and the dip ranges from 60° to 70° SE. The strongly siliceous croppings contain copper carbonates. They are rather misleading, as underground the vein appears to split up and become irregular. The tunnel follows the largest of the stringers, which ranges from 3 inches to 4 feet in width. The vein here consists of crushed granodiorite partly cemented by quartz in which there is some pyrite, copper carbonates, and chalcocite. At the face of the drift the vein is very narrow and carries no copper minerals. A north-south zone of fracture crosses the drift 120 feet out from the face. Copper minerals were not seen beyond a point 460 feet from the mouth of the tunnel and about 150 feet vertically below the surface, where there is a small pocket of ore. About 400 feet in from the portal a shoot of ore 4 feet wide and 30 feet long has been stoped for about 20 feet. In this shoot the pyrite is coated and in some places completely replaced by chalcocite (copper glance), which has suffered further alteration to the copper carbonates.

The picked carbonate ore, none of which has been shipped, though there are several hundred tons on the dump, is reported to carry between 12 and 15 per cent of copper and small quantities of silver.

Postmineral movement has taken place along the vein, forming a heavy gouge on both walls and crushing the ore to some extent.

It would seem from the short inspection of this deposit that its valuable content is almost entirely secondary and has resulted from a concentration by downward-moving solutions carrying the copper. The primary ore appears to be a noncupriferous or only slightly cupriferous pyrite, so that the source of the copper is not thoroughly clear.

Big Twenty and Mountain View system.—The Big Twenty and Mountain View veins strike N. 75° E. and stand vertical or have a steep northerly dip. At the Mountain View workings, west of the summit, there is a single, almost vertical vein which ranges from 1 foot to 2 feet in width. It is opened by three short drifts and a shaft near the base of the hill. The vein occurs in a fracture along the south side of a dark dikelike mass of granodiorite. The vein filling is largely sericitized country rock with some soft porous quartz in which there is some specularite and sulphur. In the lowest tunnel pyrite is disseminated in the altered country rock together with some specularite and a little quartz.

East of the summit on the Big Twenty ground there are three subparallel veins that seem to be without much question the continuation of the Mountain View. They outcrop near the crest of a spur that lies east and west and rises several hundred feet above the site of Granite. They are developed by several inclined shafts on the vein and by a long crosscut tunnel which could not be entered. They were seen, however, in a short crosscut midway between the croppings and the mouth of the lower crosscut. The croppings of these veins consist of iron-stained quartz from a few inches to 4 feet in width, which stands well above the surface. Underground they are represented by fracture zones, along which the granodiorite is sericitized for a width of 1 foot to 2 feet. Small iron-stained quartz stringers cut the altered rock in a few places, showing some manganese stain.

The valuable constituent of this vein is said to be gold, which occurred in small irregular pockets in the upper portion of the veins, and exploitation was not attended with success. The veins as a whole were not rich enough to work, and it seems probable that with depth they will turn into low-grade pyritic veins.

Other mines.—It is reported that at the Searchlight and Esther claims, 4½ miles south of Granite, some copper and lead ore is being mined. The Searchlight claim is situated on the west side of the summit and the copper-bearing vein is said to lie in granodiorite. At the Esther claim, on the Walker Lake drainage area, two veins are reported. One of these veins lies between granodiorite and diorite and is said to carry argentiferous galena and a little gold. The other, a copper deposit, is said to occur along a contact of diorite and limestone.

On the Nimo claim, 3 miles east-northeast of Granite, there is said to be a 2 to 4 foot vein having an easterly strike in granodiorite, near a dike of "rhyolite porphyry," which is probably a porphyritic phase of the aplite dikes common to this region. Lead-silver ores were being shipped in July, 1912, from the Flynn mine, 3 miles west of Schurz. The vein is said to run nearly east and west in granodiorite.

PINE GROVE DISTRICT

LOCATION AND ACCESSIBILITY.

The Pine Grove district (No. 13, Pl. I, p. 18) is in the northwest part of Mineral County, Nev., about 17 miles in an air line south-southeast of Yerington, the largest town in Mason Valley. The settlement of Pine Grove is 4 miles east of the Mineral-Lyon county line, on the abandoned Lobdell Summit road between East and West Walker rivers, and is shown in the lower right-hand quarter of the Wellington topographic sheet of the United States Geological Survey. The district is most easily reached from Yerington, on the Nevada Copper Belt Railroad, which joins the Southern Pacific at Wabuska and is operated by the Mason Valley Copper Co. between its mines at Ludwig and Yerington and its smelter at Thompson.

TOPOGRAPHY.

Pine Grove is on the east flank of the Smith Valley Range, which is an irregularly shaped group of mountains separating Smith Valley on the west from Mason Valley on the east. The north end of this range, in the vicinity of the Yerington copper mines, is quite narrow, but south of the West Walker River canyon it spreads out to a maximum width of 12 miles. It is roughly separated from the Sweetwater Mountains at the south end by Dalzell Canyon.

The settlement of Pine Grove stands at an elevation of 6,700 feet in the canyon north of Pine Grove Summit, whose elevation above sea level is approximately 8,500 feet. Mount Etna, a mile and a half northwest of the town, has an elevation of 7,400 feet, and the Sugarloaf, a mile and a quarter east of the town, reaches a height of 6,900 feet.

GEOLOGY.

Quartz monzonite.—The town of Pine Grove is situated in an area of light-gray coarsely granular rock. In hand specimens it seems to be feldspathic, though a little quartz is visible, together with biotite and hornblende. Under the microscope the thin sections are seen to be composed largely of feldspar with some ferromagnesian minerals and minor quartz. The feldspar is about equally distributed between orthoclase and oligoclase-andesine, but in some thin sections the plagioclase is in excess of the orthoclase, whereas in others the reverse is true. Greenish biotite is the most common ferromagnesian mineral, but hornblende is sparingly present. Magnetite, titanite, and a little apatite are common accessory minerals. The biotite is altered, some of it being bleached to a nearly colorless mica and other fragments being largely changed to chlorite. The feldspars are kaolinized, the orthoclase being altered before the plagioclase feldspars. This rock is a quartz monzonite near granodiorite, and is typical of the late Mesozoic intrusives of the Pacific coast.

This formation is cut just north of Pine Grove by a strong fault that strikes N. 60° W. and dips 40°–50° N. For 100 to 200 feet south of this fault there is a zone of more or less crushed and intensely altered quartz monzonite, which is a very dark gray to greenish-black rock. It is in this zone that the ore deposits are found.

Granite porphyry.—About half a mile east of Pine Grove, in the canyon, the quartz monzonite is intruded by a dike of reddish-gray porphyry that contains small distinct pink orthoclase phenocrysts, together with white quartz blebs and flakes of greenish biotite. Thin sections of this rock show a medium coarse granular groundmass composed of orthoclase, oligoclase-andesine, microperthite, quartz, and biotite, in which are set small, fairly well developed phenocrysts of the same minerals. The quartz and orthoclase of the groundmass are intergrown in some slides as in "graphic granite." Epidote and kaolin are secondary products from the biotite and feldspars. This rock, a granite porphyry, is probably a derivative of the quartz monzonite magma and was intruded into that rock at no great time after its formation.

Rhyolite.—Above the quartz monzonite and granite porphyry lies a series of pink to gray rhyolites. These rocks form the "Sugar Loaf," 1¼ miles east of Pine Grove. The contact lies at the west base of the peak, about 1 mile east of town. From this place it crosses the canyon at an elevation of about 5,900 feet and swings N. 73° W., forming the highest points along the summit of the ridge north of Pine Grove Canyon. These flows seem to dip to the north and northeast at low angles and are relatively thin. The lower flows are slightly porphyritic; the upper flows are more glassy. In the porphyritic flows the groundmass shows microscopic crystals of quartz and orthoclase, rather thickly scattered through the glass base. In this groundmass are a few small fragments of orthoclase, quartz, and biotite.

Basalt.—Some of the low hills in the valley 5 miles east of Pine Grove are composed of black vesicular augite basalt, which overlies the rhyolite unconformably.

ORE DEPOSITS.

History and production.—The first discoveries of gold were made on some outcroppings on the north side of the canyon at the town of Pine Grove in 1866 by William Wilson, a resident of Mason Valley. The Wilson mine covers the original location, and also about 80 acres of ground in the vicinity. For some years the district was called the Wilson, after its discoverer, but the name of Pine Grove was finally adopted from a grove of piñon trees, which the Indians visit annually to gather nuts.

In 1869, according to Raymond,[1] there were several arrastres and a 10-stamp mill in operation on oxidized ore, which ran from $30 to $90 a ton. The bullion produced at that time was said to be 0.917 fine.

The Wheeler mine, on the south side of the canyon, about three-fourths of a mile east of town, was discovered shortly after the Wilson.

In 1882 Burchard[2] reports that both the Wheeler and Wilson mines were working ores between $50 and $60 in grade in amalgamat-

[1] Raymond, R. W., Mineral resources of the States and Territories west of the Rocky Mountains, p. 116, 1869.

[2] Burchard, H. C., Report of the Director of the Mint for 1882, p. 141.

ing mills. The Wheeler mill had 15 stamps and the Wilson mill 10 stamps.

Prior to 1896 none of the ore was concentrated, and as only about 33 per cent in value of the precious metals in the sulphide ore is free milling the tailing piles contained considerable quantities of those metals. A small cyanide mill not in use in 1912 was still standing in the canyon just north of Sugar Loaf Peak. It is said that a large quantity of the tailings from the Wheeler mine had been re-treated in this mill with considerable success.

During the later years of development in the mine the low-grade material averaged 0.28 ounce in gold to the ton and the high-grade sulphide ore 3.4 ounces gold, 0.3 ounce silver, and 0.24 per cent copper to the ton. It is estimated by Mr. Deleray, superintendent of the Wilson mine, that the production from that property between its discovery and 1893 amounted to about $5,000,000 and that about $3,000,000 was taken from the Wheeler mine during the same period. Since 1893 the mines have not been worked continuously, but in 1909 the Wilson mine was purchased by the Pine Grove Nevada Gold Mining Co., so there is some hope that the camp may again become active. According to the mining journals the Wilson mine has been reopened and equipped with electrically driven machines.

The following table of production of the Pine Grove district is taken from the mineral resources reports of the United States Geological Survey for the years 1902 to 1911, inclusive. The value of ore a ton according to these figures ranges from $6 to $22, the average being $8.65.

Production of gold and silver from the Pine Grove district, Mineral County, Nev., from 1902 to 1911, inclusive.

Year.	Crude ore treated.	Gold.	Silver.	Total value.
	Tons.		*Ounces.*	
1902	1,105	$11,270	5,917	$14,164
1903	2,675	63,000	28,584	75,000
1904	2,300	14,238	1,882	15,310
1905	150	1,230	242	1,376
1906	1,231	11,090	1,821	12,310
1907	9,735	12,393	3,264	14,547
1908	632	14,143	277	14,310
1909	.....	.....	.....	.....
1910	273	3,246	2,817	4,767
1911	1,024	11,914	81	11,957
	19,125	142,524	44,885	163,741

Development.—The Wilson mine is developed by a series of tunnels that join and cross one another in a most intricate manner. From the tunnels there are many long-filled stopes, shoots, and galleries, making a network of openings in the ore zone which are estimated to total about 3 miles of workings. All this work reaches a maximum depth of 140 feet below the outcrop. A 300-foot incline shaft is located near the mouth of the main working tunnel, but its mouth was caved and water was standing within 100 feet of the surface, so it could not be entered. Most of this work was done by lessees, as the mine was largely worked under that system. The lessees paid the company a royalty of 50 per cent of the net returns on all ore produced.

The ground is heavy, necessitating the use of large timbers, and a great many of the old stopes are either partly or completely filled with waste, many of them being caved.

A long crosscut tunnel has been started to cut the ore body 500 feet below the outcrop, but as yet has been driven only through the overlying rhyolites. This tunnel has its mouth in the first canyon north of Pine Grove Canyon, about 1½ miles north-northeast of town. It is said that there are about 4 miner's inches of water a minute developed by this tunnel, which would be an adequate supply for a mill located at the tunnel mouth.

The Wheeler mine is developed by three irregular tunnels and connecting stopes, raises, and chutes, through a vertical distance of about 120 feet. The work is in rather bad condition, owing to its long idleness. Most of the stopes are filled, but the ore is visible at many places in the drifts.

Occurrence of the ore.—The ore bodies in the Wilson and Wheeler mines lie in a zone of intense crushing and alteration immediately south of the well-marked fault, striking N. 60° W., that cuts the quartz monzonite. This fault dips 40°–45° N. and is marked by a heavy clay gouge. Rounded pebbles of the altered rock are present in this gouge, which seems to be largely iron-stained clay, together with much sericite. Calcite crystals are developed in it, some of them 2 inches across. The gouge ranges from 4 to 6 feet in thickness and slickensides in it are all about parallel to the dip of the vein. The direction of movement was not entirely clear, but the faulting appears to have been normal.

The quartz monzonite for 150 to 160 feet south of the fault has been crushed and altered to a high degree. In this zone there is a considerable quantity of disseminated pyrite and a large number of small interlacing quartz stringers carrying pyrite that range from one-eighth inch to 3 inches in width. There are also lenses of quartz and pyrite that may be 2 or 3 feet thick extending along the strike for 10 to 150 feet. These lenses of sulphides constitute the good ore, which is said to have been of better grade in those portions from 8 to 10 inches thick. They have a very flat dip to the north, between 10° and 15°, so that many pillars were left in the mining operations. The lenses overlap both in strike and dip, and there was apparently no indication to direct new drifting at the end of a lens. There are a few postmineral faults about parallel to the main fracture that are vertical and have moved the rock on the north side downward from 6 to 8 feet.

Character of the ore.—The small streaks and large lenses of quartz and sulphides constitute the ore. The sulphide is 95 per cent pyrite, though a little chalcopyrite occurs in places, particularly in narrow veinlets branching off from the lenses. The ores are all somewhat oxidized to a depth of 170 feet. Limonite is common and locally there are small areas of copper-stained ore where the chalcopyrite was most abundant.

The valuable constituents of the ore are gold and silver, and it is said that at the Wilson mine, at the west end of the mineralized area, the ratio of gold to silver was higher than at the Wheeler mine east of Pine Grove. The bullion from the Wilson mine is said to have sold for $18.75 an ounce, whereas that from the Wheeler mine was worth only $15 an ounce. Above the 170-foot level about 50 per cent of the gold was free in the semioxidized ores and at the surface very rich free-milling ore bodies were found. It is said that at a depth of 250 feet in the Wilson mine there is practically no oxidation of the sulphides, which have a value of about $10 a ton.

Alteration of the quartz monzonite.—For about 200 feet south of the main fault, and for an unknown distance north of it, the normal light-gray coarsely crystalline quartz monzonite is crushed and altered to a dark-gray, almost black rock. The alteration is progressive, becoming less and less intense as the fault is left. The beginning of the alteration is shown in the slightly changed rock by a sericitization of the oligoclase-albite and by a partial alteration of the original green hornblende to brownish-green biotite. A small amount of epidote is also seen. In the ore zone most of the plagioclase feldspar is replaced by greenish-brown biotite and sericite, and all the original hornblende is changed to the dark mica. In this rock there is widely disseminated pyrite and numerous interlacing quartz stringers cut the formation in all directions.

At the fault the rock is completely altered to an aggregate of quartz and brownish-green biotite, together with a little orthoclase, apparently left from the original rock. All the plagioclase and hornblende are gone, but the magnetite seen appears to have escaped alteration. This rock contains abundant grains of disseminated pyrite as well as veinlets of quartz and pyrite. Calcite is developed to a small extent in veinlets in the ore zone and forms thin crusts coating the joint planes of the rock.

This alteration must have been accomplished at depth by hot ascending solutions rich in potassium. The ores were formed as replacements of the crushed and altered rock by the same solutions that caused the alteration.

Biotitization of the wall rock by vein-forming solutions, though not of wide occurrence, has been noted at a number of places. Lindgren[1] found biotite—

replacing hornblende and feldspars * * * in veins carrying tourmaline (Meadow Lake, Cal.); replacing the same minerals it appears abundantly in the gold-copper veins of Rossland, B. C. A greenish mica, probably biotite, occurs, replacing quartz, in small veinlets, associated with quartz, garnet, tourmaline, actinolite, and zinc blende, in the Bunker Hill and Sullivan mine, Idaho.

Hatch[2] reports that—

in the near neighborhood of the quartz lodes [at Kolar, India] a characteristic brown mica is abundantly developed * * * genetically connected with the mineralization of the lodes, whether by vapors from below or by ascending mineralizing solutions.

In a footnote he further suggests that the "brown mica has been produced by deep-seated vapors attacking the hornblende and supplying the requisite amount of water and alkalies."

At Bingham, Utah, Boutwell[3] notes the alteration of quartz monzonite in the vicinity of zones of strong shattering. He says:

Conspicuous areas of granular quartz are numerous, the orthoclase is highly sericitized, and the femic minerals are represented by numerous irregular patches of small individuals or flakes of dense brown biotite. The quartz and sericite are clearly secondary, and though no direct proof of the age of the biotite has been found it resembles secondary biotite and may be secondary also. Magnetite, excepting occasional grains, has disappeared, and large amounts of chalcopyrite and pyrite are present in the form of rounded grains, chains, and veinlets embedded in secondary quartz, flaky biotite, and sericitized feldspar.

In the gold veins of Dahlonega, Ga.,[4] which "lie on the contact of either mica schist and amphibolite or of mica schist and granite," Lindgren mentions garnet, apatite, ilmenite, muscovite, dark-green mica, and green hornblende in the altered wall rocks of the veins, which contain very few striated feldspar grains, though much feldspar is present, and most of it, as indicated by its optical properties, is an oligoclase, although some of it may be albite. In the same veins garnet, hornblende, apatite, and a green mica may develop along the quartz veins. He concludes that "In the prevailing class of Dahlonega mines the products of alteration are such minerals as occur in areas of regional metamorphism or in contact zones."

In his report on the Juneau gold belt of Alaska, Spencer[5] has described the alteration of the diorite, particularly where this rock

[1] Lindgren, Waldemar, Metasomatic processes in fissure veins: Am. Inst. Min. Eng. Trans., vol. 30, pp. 609–610, 644–645, 1901.

[2] Hatch, F. H., The Kolar gold field, India: India Geol. Survey Mem., vol. 33, p. 7, 1902.

[3] Boutwell, J. M., Economic geology of the Bingham mining district, Utah: U. S. Geol. Survey Prof. Paper 38, p. 168, 1905.

[4] Lindgren, Waldemar, Notes on the Dahlonega mines: U. S. Geol. Survey Bull. 293, pp. 120–122, 1906.

[5] Spencer, A. C., The Juneau gold belt, Alaska: U. S. Geol. Survey Bull. 287, pp. 62–63, 1906.

contains the greatest number of quartz veins, into a fine-grained substance, composed mainly of quartz, calcite, brown mica, and chlorite, by the addition of potash.

Knopf[1] has noted a similar alteration of the igneous rocks in the northern end of the Juneau belt by a large introduction of albite, the conversion of the hornblende and other amphiboles into biotite, and the introduction of apatite. This alteration shows a large addition of soda and potash demanded by the formation of albite and biotite, and a heavy loss of magnesia, lime, and iron. Knopf concludes that the mineralizing solutions were rich in soda and potash and that they were hot, ascending waters of deep-seated origin.

Spencer[2] has recently found a similar alteration of quartz monzonite in the Ely district, Nev. He finds the most intense alteration along zones of crushing and concludes that the alteration was produced by hot aqueous solutions carrying soda and potash which have changed the hornblende, plagioclase, and magnetite of the original rock into sericite and biotite and have deposited pyrite, chalcopyrite, calcite, and possibly quartz.

Rockland mine.—The Rockland mine, situated about 3 miles southeast of Pine Grove, though not visited, is presumably of the same type of ore deposits as the Wheeler and Wilson mines and may be on a continuation of the same fault zone, though it seems more probable that it lies in another zone of fracture. The "vein" is said to strike northwest and to dip 45°–55° NE. The footwall is the dark altered quartz monzonite and the hanging wall a "light-colored porphyry," presumably the intrusive granite porphyry, though it may be the rhyolite. The property is developed by three drift tunnels. The ore is said to contain very little copper, the value being chiefly gold and silver, the bullion having a value of $15 an ounce.

There is said to be a 20-stamp amalgamation mill on the property, lately installed to replace a dry process mill, which was unsuccessful on account of the large quantity of clay in the ore.

FUTURE OF THE DISTRICT.

This type of ore has probably been formed at considerable depth, for the accompanying alteration, as pointed out by Lindgren,[3] is such as to preclude the theory of shallow deposition. It seems probable, therefore, that the mineralization along the fault zone will continue for a considerable distance below the surface. It is to be expected, and is already shown in the lower workings of the Wilson, that the

[1] Knopf, Adolph, The Eagle River region, southeastern Alaska: U. S. Geol. Survey Bull. 502, pp. 36–41, 1912.

[2] Unpublished manuscript.

[3] Lindgren, Waldemar, Metasomatic processes in fissure veins: Am. Inst. Min. Eng. Trans., vol. 30, pp. 609–645, 1901.

grade of the ore will not be as high as at the surface, but the unaltered sulphides at a depth of 250 feet carry $10 a ton, and ore of even lower grade has been successfully worked at numerous places. This pyritic ore, with such a small proportion of copper minerals, is amenable to cyanide treatment, and it seems entirely possible that if a sufficient quantity of ore can be treated the properties could again be producers. In fact, reports in the mining journals indicate that these mines are to resume production in the near future.

AURORA (ESMERALDA) DISTRICT, MINERAL COUNTY, NEV.

LOCATION AND ACCESSIBILITY.

The old Esmeralda mining district at Aurora, Nev. (No. 14, Pl. I, p. 18), is 28 miles in an air line southwest of Thorne, a town on the Hazen-Tonopah branch of the Southern Pacific and its nearest railroad point. The town of Aurora is 3 miles east of the California-Nevada boundary, 16 miles north of Mono Lake and 1½ miles east of Bodie Canyon. The region is shown near the center of the west side of the Hawthorne topographic sheet of the United States Geological Survey.

The district is most easily reached by the automobile stage which runs daily between Hawthorne, Nev., and Bodie, Cal. It is possible to enter this part of Nevada by way of Minden, the southern terminus of the Virginia & Truckee Railway. Stages operate between that town and Wellington and thence south to Bodie, Cal., but the trip requires three days in contrast to the half-day run from Hawthorne.

HISTORY AND PRODUCTION.

The Old Esmeralda, near the southern limit of the productive area (see Pl. XVI, *A*, p. 148), was the first vein discovered in the Aurora district. According to Wasson,[1] James M. Brawley, J. M. Cory, and E. R. Hicks made the discovery on August 22, 1860, and immediately located four claims. The town of Esmeralda was built in the gulch just east of the discovery, but later in the year the present town site of Aurora, 1½ miles north, was laid out. The first mill, owned by Edmund Green, was put in operation in 1861, and was followed shortly by several arrastres and mills. In 1864 there were 17 amalgamation mills in the district, the largest, which had 30 stamps, being the Real Del Monte in Bodie Canyon. Up to the year 1864 the camp was very prosperous. Aurora had a population of about 10,000 and was the county seat of Mono County, Cal. During the year 1864, however, misfortunes befell the camp. The California-Nevada boundary was run and showed that the Esmeralda district lay in Nevada;

[1] Wasson, Joseph, Bodie and Esmeralda: a pamphlet published in 1878 by the Mining and Scientific Press, San Francisco, Cal.

the rich bonanzas in the Wide West vein on Last Chance Hill became exhausted and bitter litigation over the ownership of the veins on Last Chance Hill developed. The camp, however, continued to prosper until 1882, though the supply of $75 ore, which in earlier times could not be mined, was then becoming depleted. In 1880 an English company acquired possession of the main group of claims on Last Chance Hill. It began operations in 1887, starting the Real Del Monte shaft and connections with the Durant vein on Middle Hill, but suspended work in 1892 after a vain effort to keep the lower workings of the 800-foot shaft free from water.

Most of the claims in the Esmeralda district were owned in July, 1913, by two companies, the Cain Consolidated Co. and the Aurora Mines Co. The Aurora Mines Co.'s chief group, containing 11 claims, lies on Silver Hill, though they own 5 claims on Aurora Hill. The Cain Consolidated Co. controls about 40 claims, among which are some of the famous producers of the district. In the summer of 1912 these holdings were under option to certain financiers of Tonopah, Nev., who have, according to reports of the mining journals, taken up the ground and started operations.

Most of the productive ground of the district has now been acquired by the Goldfield Consolidated Mining Co. A 500-ton cyanide mill has been built, and there is every prospect that Aurora will again be a producing camp.

The records of production are incomplete. According to a statement of Wells, Fargo & Co. the bullion shipped through them up to 1869 had a value of $27,000,000. Mr. Wasson[1] gives the following table of gold bullion shipped without insurance:

Bullion shipped from Aurora without insurance from 1861 to 1869, inclusive.

1861	$43, 417. 28	1867	$130, 656. 89
1862	173, 148. 82	1868	98, 188. 88
1863	546, 019. 16	1869	28, 166. 50
1864	952, 023. 29		
1865	237, 185. 23		2, 365, 968. 82
1866	158, 162. 77		

He further says that between seven and eight million dollars' worth of bullion was shipped by express in 1864 and about $12,000,000 prior to the year 1869. If the reports of production of some of the stopes are taken into consideration, even so large a sum as $27,000,000 seems a small showing for the camp.

TOPOGRAPHY.

There are four rather low hills south and east of the town of Aurora (see Pl. XIV), known as Silver, Middle, Last Chance,

[1] Wasson, Joseph, op. cit.

and Humboldt. The town has an elevation of 7,415 feet above sea level. Silver and Middle hills are separated by Esmeralda Gulch. They are long, northward-sloping spurs from the Brawley Peaks, which rise to a height of 9,557 feet about 2¼ miles south of the town. Last Chance Hill, east of Aurora, is a low divide, less than 150 feet above the valley, which separates Willow Creek from the Gregory Flat drainage basin. Humboldt Hill, a low rounded knob about three-fourths of a mile northeast of Aurora, rises to a height of a little over 7,600 feet. The mines of the Aurora district are located on these four hills, though at the east end of the flat north of town and about 150 feet higher there are a few veins near Humboldt and Last Chance hills.

GEOLOGY.

CHARACTER AND DISTRIBUTION OF THE ROCKS.

The rocks exposed in the Esmeralda district are, with a single exception, of volcanic origin. In the bottom of Willow Gulch, about 2 miles southwest of Aurora (see Pl. XIV), there is a small, indistinct exposure of a rock that appears to be the basement on which the flows were extruded. It is a coarsely porphyritic, granular rock, and is probably to be correlated with the granodiorite and associated rocks of the Sierra Nevada. At least three series of flows overlie this granular rock. The oldest of these flows consists of grayish-green altered rocks that are largely biotite-quartz latites, together with some andesites. This series is exposed on Silver, Middle, Last Chance, and Humboldt hills, and extends southeast up Willow Gulch for an unknown distance. These rocks, which are at least 900 feet thick, inclosed all the veins of the district, none being found in the younger rocks.

Above these flows of intermediate chemical composition lies a series of light-gray to brownish-gray rhyolites that are particularly well exposed on the flats north and northwest of Aurora along Bodie Creek, where they are 1,000 feet thick, and also on Granite Mountain, about 1 mile southeast of town, where there is a remnant of the series about 300 feet thick. Above both the andesites and rhyolites lies a black vesicular basalt that forms Aurora Crater (see Pl. XIV) and covers a large expanse of country to the west of Granite Mountain. It ranges from about 10 to over 600 feet in thickness. Its weathered surfaces are brown.

All these flows appear to have a gentle dip to the north-northwest. It seems probable that there was a time of erosion between the andesite and rhyolite eruptions, as the rhyolite flows rest on an uneven surface that looks like an erosion surface.

There was unquestionably an interval of considerable length between the rhyolite and basalt eruptions, for the base of the later

U S GEOLOGICAL SURVEY
GEORGE OTIS SMITH, DIRECTOR

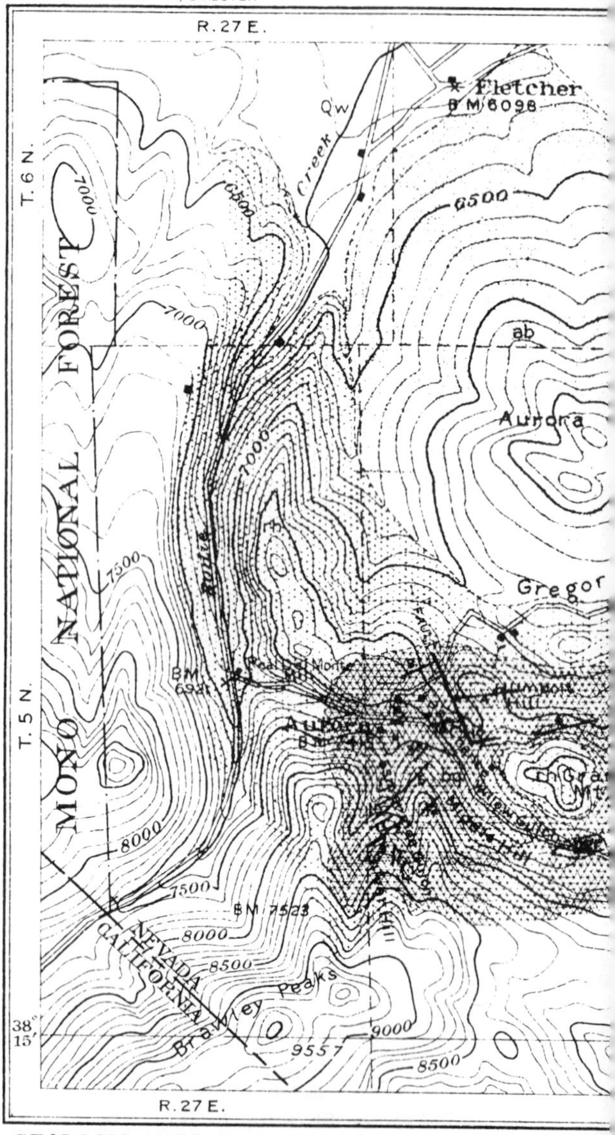

GEOLOGIC SKETCH MAP OF THE VICINITY OF AU

Base from topographic map of the Haw

1 ½ 0 1 2

Contour interval 100 f
Datum is mean sea le
1914

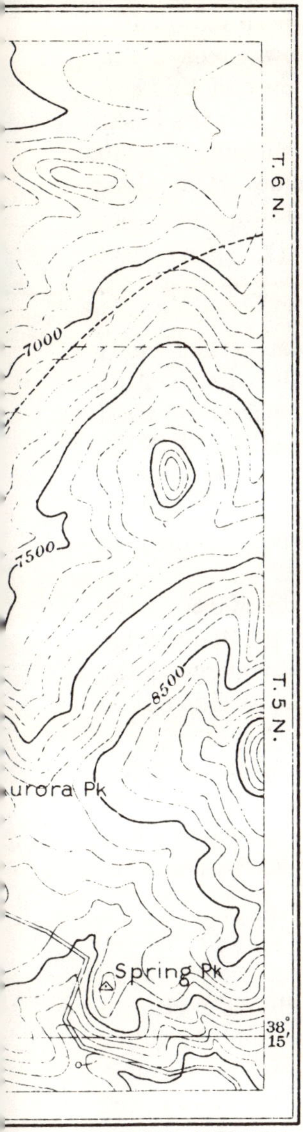

ERAL COUNTY, NEVADA

ngle

4 Miles

LEGEND

Qw
Quaternary wash

ab
Augite basalt
(*Cap rock*)

rh
Rhyolite
tuffs and flows

bq
Biotite-quartz latite
and associated andesite
porphyries

Porphyritic granite

Veins
(*Numbers refer to list of names*)

Fault

LIST OF VEINS

1 Antelope
2 Bald Eagle-Spotted Tiger
3 Durant
4 Humboldt-Silver Lining
5 Live Yankee
6 Martinez-Juanita
7 Old Esmeralda
8 Prospectus
9 Radical
10 Real Del Monte
11 Sonora
12 Summit
13 Utah-Cortez
14 Wide West-Last Chance

flows rests in the bottoms of gulches in some places and on the tops of ridges at other places.

The latites and associated andesites seem to have been exposed by the erosion of the capping rhyolite along the Willow Creek drainage basin, and Granite Mountain seems to be a remnant of this capping which escaped erosion. It does not seem probable that the basalt ever extended much beyond its present limits, as shown on Plate XIV, for the edges of the flows are fresh and in some places along the gulch northeast of Gregory Flats show the piled-up, overturned marginal portions of quickly cooled lava sheets.

TOPOGRAPHIC EXPRESSION OF THE DIFFERENT ROCKS.

The oldest flow rocks are all much altered and are rather easily eroded, except where they have been silicified near the veins. As a consequence the mineralized hills have, as a rule, even and rather gentle slopes, as is shown in the view of Last Chance Hill (Pl. XV). At the southern end of Silver Hill, where the rock is much silicified near the Bald Eagle, Spotted Tiger, and Radical veins, the generally smooth andesite surfaces are interrupted by steep cliffs, as shown in Plate XVI, *A*.

The rhyolite series weathers in rough cliffs and the surfaces of the flows are marked by small steep-sided gullies. On the long ridge northwest of Aurora, near Bodie Creek, the topography suggests the badland forms at many points, especially on the northeast side of the ridge.

The surface of the basalt flows is very rough, making the crossing of these areas difficult, even on foot. Aurora Crater is a basaltic vent, the northwest rim of which has been cut through by erosion. It is a beautiful example of a small volcano, with the successive flows clearly traceable on its rough scarred sides.

PETROGRAPHY.

Porphyritic granite.—The single exposure of porphyritic granite in the district lies in the bottom of Willow Gulch, about 1¼ miles southwest of Aurora. Its boundaries are not well shown on account of the wash, but the andesitic flows clearly rest on this basement. The outcrop is small and deeply weathered, practically no fresh rock being visible. Numerous large pink orthoclase crystals, the maximum length being 2 inches, are present in the residual sand covering part of the area. The weathered surfaces have a light greenish-gray color, owing to the alteration of the constituents. The rock is rather coarsely granular throughout and contains very large, zonally built, pink orthoclase phenocrysts.

LAST CHANCE HILL, AURORA, MINERAL COUNTY, NEV.

Shows gentle slopes of andesitic flows.

In thin sections the groundmass of this coarse porphyry is seen to be inequigranular. None of the minerals, except the phenocrysts and accessory minerals, show any crystal form. The minerals present in this rock, named in the order of their abundance, are orthoclase, quartz, microperthite, green hornblende, brown biotite, microcline, muscovite, and oligoclase. The accessory minerals are titanite, magnetite, and apatite. The ferromagnesian minerals are somewhat chloritized, and the feldspars are more or less kaolinized. Some of the muscovite appears to be primary, but part, at least, is bleached biotite. The titanite and magnetite are closely associated and intergrowths of these two minerals are common.

Biotite-quartz latite.—The general country rock on the hills in the vicinity of Aurora, in which the veins are found, is a greenish-gray to gray altered porphyry which ranges from rather fine to medium grain. Few of the phenocrysts are more than an eighth of an inch in diameter, and most of them are less. The most widely distributed type of rock has a fine-grained greenish-gray groundmass, thickly studded with small white lath-shaped phenocrysts. All the rock of this type carries some disseminated pyrite, which is particularly abundant near the veins.

The thin sections show that this rock originally consisted of phenocrysts of andesine, biotite, and possibly pyroxene, set in a fine-grained matrix of andesine, with some ferromagnesian minerals. Small interstices of the groundmass contain intergrowths of quartz and orthoclase. The rock is a biotite-quartz latite. All the rocks are very much altered, presumably by the hot calcareous solutions which deposited the veins. The andesine phenocrysts are altered to calcite, sericite, some quartz, and some of them show a little green epidote. The ferromagnesian minerals are completely altered to chlorite and some magnetite. The groundmass is altered to an aggregate of sericite, chlorite, and quartz. Near the veins the alteration has been much more intense than at a distance of 150 to 200 feet from them. In these highly altered zones quartz has been added to the body of the rock, which is also cut by stringers of quartz and calcite. Sericite and epidote are also much more abundant in the rock near the veins, whereas chlorite is more commonly developed in the rock at a distance from the veins.

Andesite.—Near the bottom of Esmeralda Gulch, 1 mile south of Aurora, there is a fine-grained light-green porphyry apparently intrusive into the biotite-quartz latite, though it may be an underlying flow. This fine-grained dark rock is exposed in several other localities in the district, and it is probably rather widespread in distribution.

This rock is much altered and contains disseminated pyrite in small quantities and is cut by quartz and calcite stringers.

Thin sections of this rock show that its groundmass is composed of microscopic lath-shaped crystals of andesine and augite, the latter mineral altered to chlorite. In this groundmass are set small well-developed phenocrysts of zonally built andesine and of augite, both of which are altered, the augite to green chlorite and the feldspars to grayish aggregates of sericite and chlorite.

Rhyolites.—The rhyolite series is made up of a number of relatively thin flows, all of which are glassy. They range in color from gray through green to purple. Some of them appear to be tuffaceous, but the majority are typical flow rhyolites. On the flat north of Aurora some pearl-gray perlitic rhyolites are seen near the top of the series. Flakes of biotite are seen in all of these rocks, and quartz can usually be detected with the unaided eye. Thin sections show that the rock consists of a glassy base having, as a rule, distinct flow structure, which contains a few phenocrysts of quartz, orthoclase, and biotite. Some of the slides show that the groundmass suffered some devitrification, accompanied by the development of chlorite and sericite.

About one-fourth of a mile southeast of the Old Esmeralda Tunnel a small, indistinct body of rhyolite has been altered to a soft white mass by hydrothermal action, but the flows at other places show no alteration by hot waters.

Basalt.—The basalt of Aurora Crater is a very fresh vesicular black rock showing a few small green olivine crystals to the unaided eye. Under the microscope the groundmass is seen to be composed of microscopic labradorite laths and grains of nearly colorless augite set in a black glass paste. The flow structure is well shown by the rough parallel orientation of the long dimensions of the plagioclase laths, many of which bend around the vesicular openings.

QUATERNARY GRAVELS.

The Quaternary deposits on lower Bodie Creek, shown at the top of Plate XIV, consist of unconsolidated sands, gravels, and silts, which a little north of the area shown on the map are quite thick and extensive. The surface is covered by fine sandy loam, which under irrigation has produced excellent crops.

On Last Chance Hill there is a small area underlain by roughly stratified volcanic material, shown in the caved Chihuahua stope. (See Pl. XVI, *B*, p. 148.) This material ranges from a few feet to a maximum of 20 feet in thickness and appears to have been reworked by streams.

In the canyon northwest of Gregory Flats and about 1½ miles due north of Aurora a warm spring issues from beneath the basalt of Aurora Crater. This spring deposits limonite and aragonite. The aragonite forms crusts from one-fourth inch to 4 inches in thickness, though most of the crusts are less than 2 inches thick. The entire deposit covers an area about 150 feet square to a maximum depth of 15 feet.

ORE DEPOSITS.

Distribution of the veins.—The ore deposits of the Esmeralda district occur as veins that cut the biotite-quartz latite and associated andesite over an area extending in a northeast-southwest direction, about 2 miles in length by 1¼ miles in width. Aurora is near the center of the northwest side of the productive area. On Silver and Middle hills the veins are rather closely spaced and have, with one exception, a persistent strike of about N. 45° E., though strikes between N. 40° E. and N. 50° E. are seen in many places, even along veins whose average course is N. 45° E. These veins all dip to the southeast but at different angles. The largest and apparently the strongest veins—that is, the Eureka, the Antelope and Lady Jane, the Cortez and Utah, and the Spotted Tiger and Bald Eagle veins—all dip between 45° and 60° SE. into the hill, but some of the smaller veins slope southeast at much flatter angles. An exceptional vein system on Silver Hill is represented by the Old Esmeralda and Radical veins, which strike about N. 10° E. and stand nearly vertical. The outcrops of these veins are wider than any of the northeast-southwest system, the Old Esmeralda being 60 feet wide and the Radical between 20 and 30 feet in maximum width.

On Last Chance and Humboldt hills the veins strike more nearly east and west, ranging between N. 60° E. and N. 80° E., and with the exception of the Humboldt and Prospectus veins, which dip 80° N., they dip to the west-southwest at angles ranging between 65° and 75°. Some of the smaller veins have flatter dips, but the strong, well-defined ones stand more nearly vertical. The veins on these two hills have been displaced by a nearly vertical fault, whose strike ranges from N. 20° E. at the Humboldt vein to N. 30° W. at the Real Del Monte vein. The horizontal displacement along this fault has amounted to over 600 feet, the veins west of the fault being at least that far north of their continuation on the eastern side. Thus the Humboldt vein on the east side of the fault along the crest of Humboldt Hill is the Prospectus west of the fault. (See Pl. XIV, p. 142.)

It is said that about 6 miles northeast of Aurora lies a small area of andesite in which there are some veins, but the locality was not visited.

Character of the veins.—The veins of the Esmeralda district range from a fraction of an inch to 70 or even 80 feet in width. They are as a rule between 18 inches and 4 feet wide, and can generally be traced for several hundred feet along the strike. They are not simple, clean-cut veins, but send off numerous small interlacing branches into the walls, particularly on the footwall side. This tendency of the veins is well seen on the edge of the Old Chihuahua stope on Last Chance Hill. (See Pl. XVI, *B*.)

Along some of the veins there has been postmineral movement. This movement has usually taken place along the hanging wall and has been slight, producing in most places a thin clay parting between the country rock and quartz.

The veins consist in great part of finely granular, white, barren-looking quartz. In some places the quartz is so fine grained that it has a milky-white porcelain-like appearance. The veins are banded by crustification, the different bands being due to the difference in size of the quartz grains. In all the veins there are small druses lined with minute clear quartz crystals. The rich ore is always marked by irregular wavy streaks of what appears to be dark quartz, cutting the white low-grade or barren vein filling. (See Pl. XVII, *A*.) In reality these rich streaks are made up of quartz, adularia, argentiferous tetrahedrite, and small amounts of pyrite and chalcopyrite, together with a soft bluish-gray mineral supposed to be a combination of gold and possibly silver with selenium. Some free gold is found here and there in the richest ore now mined, and the old stopes are said to have contained large quantities of free gold.

The adularia is notably absent from most of the white barren quartz, but was found in soft, narrow kaolinized bands in white quartz ore from the Humboldt shaft. In practically every thin section of the ores studied there is a small amount of sericite in thin flakes cutting the quartz crystals. In all the thin sections of very rich ore adularia is abundant, being associated with the quartz and commonly inclosed in the interlocking quartz grains. As a rule the rhombic forms of adularia are not seen, the mineral occurring in irregular masses.

Qualitative tests of a small piece of rich ore from the 350-foot level of the Durant vein at Aurora show the presence of selenium but no tellurium. It also contains a rather large quantity of iron and copper, smaller quantities of silver and gold, and some antimony. A polished section and thin section of this ore show the undoubted presence of pyrite, chalcopyrite, tetrahedrite, and free gold. There is also a small quantity of a soft bluish-gray mineral, that is distinct from the tetrahedrite, which is thought to be a selenium-gold and possibly silver compound. This mineral occurs in minute specks and could not be separated from the other constituents.

A. SOUTH END OF SILVER HILL, AURORA, MINERAL COUNTY, NEV.

Shows croppings of Old Esmeralda vein and cliffs where andesitic flows are silicified at the junction of several veins.

B. CHIHUAHUA STOPE ON LAST CHANCE HILL, MINERAL COUNTY, NEV.

Branching veins in footwall of main vein at left and stratified volcanic material in right center.

A. STREAK OF RICH ORE FROM SPOTTED TIGER VEIN.

Natural size.

B. ORE FROM A SMALL VEIN CUT BY THE MONARCH TUNNEL.

Shows typical quartz-adularia mineralization after calcite.

The presence of selenium without tellurium sets these veins apart, for there are only a very few mining districts in the United States where this combination of ores is found.

Spurr[1] says that at Tonopah the veins are chiefly due to replacement of the andesite by quartz and the ore minerals along zones of fracture. Crustified veins clearly due to filling of open spaces are exceptional at Tonopah. The mineralizing agent he considers to be "volcanic waters that were hot and ascending." The primary ores at Tonopah, according to Spurr, contained quartz, adularia, carbonates of lime, iron, magnesia and manganese, silver sulphite, probably polybasite or stephanite, and argentite, silver chloride, chalcopyrite, pyrite, galena, sphalerite, and gold in an undetermined form, and silver selenide.

At Republic the veins which seem to bear more resemblance to those at Aurora, according to Umpleby,[2] occur along fissure fillings that have an average width of 3½ feet. The unaltered vein material is a firm white quartz with wavy ribbons of a bluish-gray cast. The veins are made up of quartz, chalcedony, opal, calcite, and adularia, carrying inconspicuous amounts of pyrite and chalcopyrite, with silver and possibly gold, in association with antimony, sulphur, and selenium. The most striking feature of the Republic ores is the extremely barren appearance of the quartz. Fluorite was noted in the slides. The silver is thought to be partly in the form of silver selenide and partly as a component of gray copper. Some gold is free, but most of it is probably combined with selenium and tellurium.

Lindgren[3] says of the Republic veins that the banding is due to the difference in size of the quartz grains; that the ore minerals occur in extremely fine distribution in thin black streaks, generally near the walls. In the rich portion of the veins tetrahedrite and chalcopyrite have been identified, but the principal ore mineral, presumably a selenide of gold and silver, occurs in such fine distribution that it has not yet been isolated.

As it was impossible, on account of the condition of most of the mine workings, to study the veins at Aurora at depth, a satisfactory understanding of the distribution of the good and poor ore was not reached during this reconnaissance. From what could be learned, however, it would seem that the rich ore occurred in relatively small

---

[1] Spurr, J. E., Geology of the Tonopah mining district, Nev.: U. S. Geol. Survey Prof. Paper 42, pp. 83–104, 1905.

[2] Umpleby, J. B., Geology and ore deposits of the Republic mining district: Washington Geol. Survey Bull. 1, p. 37, 1910.

[3] Bancroft, Howland, The ore deposits of northeastern Washington: U. S. Geol. Survey Bull. 550, 1914.

shoots in the large barren veins. There were five such shoots, which were exceptionally large on the Wide West vein on Last Chance Hill.

In the majority of the veins it is understood that the rich ore streaks, ranging from a fraction of an inch to 6 inches in width, were as a rule found near the walls, particularly the hanging wall, and that they were not continuous along the veins for any considerable distance. It is almost certain, however, that the barren-looking white quartz, where it shows even a slight suggestion of the bluish color, carries gold.

The veins are said to have been in general considerably wider in the richer portions. This was particularly the case on the Wide West vein, where some of the stopes were as much as 60 feet wide, though the leaner portion of the vein between the stopes ranged from 6 to 10 feet in width.

As only the surface workings and outcrops of these deposits could be studied it is not possible to give detailed descriptions of the veins.

Tenor of the ore.—The average ore from any vein is probably of low grade. Ore which is taken from rich shoots may run up to $1,000 a ton. The average gross value of the ore is reported to be about $6 to $8 a ton, the ratio of gold to silver being 1 to 2 or 1 to 5. It is said that in the ore mined in the early days from the rich stopes on Last Chance Hill the ratio of gold to silver was as 4 to 2.

Origin of the veins.—The veins of the Esmeralda district were formed in open fissures by hot ascending, very siliceous solutions. These solutions were capable of altering the inclosing biotite-quartz latite for considerable distances from the veins. In one or two small veins cut by the Monarch tunnel on Silver Hill the quartz shows the typical form of replacement after calcite, common to the quartz-adularia type of veins (Pl. XVII, *B*). Pure white calcite was found on the dump of the Humboldt shaft that was said to come from the 450-foot level, though none was found above the 100-foot level in the mine. This ore seems to show that calcite was deposited before the quartz, which appears to replace the carbonate. It is questionable, however, if any considerable time intervened between the deposition of the calcite and the entrance of the silica-bearing solutions. Certainly in most of the ore there is little suggestion that the quartz is secondary after calcite, except that none of the quartz shows the crystal forms of this mineral usually seen where it is deposited alone in open fissures.

The age of the formation of these veins is not certainly known, though they were formed after the eruption of the biotite-quartz latite and associated andesite and before the succeeding flows of rhyolite. It is probable that they are representatives of the late Tertiary mineralization common to the Great Basin region.

HAWTHORNE DISTRICT, MINERAL COUNTY, NEV.

LOCATION AND ACCESSIBILITY.

The Hawthorne district comprises a large area, tributary to the town of that name, on the east side of the Wassuk (Walker River) Range and on the northwest side of the Excelsior Mountains. The town of Hawthorne is situated at the south end of Walker Lake, 7 miles south-southwest of Thorne, a station on the Hazen-Tonopah branch of the Southern Pacific Railroad. The region is shown in the west-central part of the Hawthorne topographic sheet of the United States Geological Survey. The mines visited by the writer were in the immediate vicinity of the town of Lucky Boy, 3 miles southwest of Hawthorne in an air line. (See No. 15, Pl. I, p. 18.)

PRODUCTION.

The figures showing the production for this region are not accurate, as in some years the production of the mines at Buckley Camp and Garfield was combined with that of the mines near Hawthorne. The Survey has recorded a total production since 1904 from the district of $1,090,867, extracted from 17,782 tons of crude ore. From 1907, when the veins near Lucky Boy were discovered, to 1911, inclusive, 13,968 tons of ore have been mined from these veins, carrying $21,387 in gold, 1,770,279 ounces of silver, 54,206 pounds of copper, and 2,982,041 pounds of lead, having a total value of $1,076,235.

TOPOGRAPHY.

Hawthorne, which lies at an elevation of 4,326 feet above sea level, is situated near the center of the flats at the south end of Walker Lake. About 4 miles west of the town the Wassuk Range rises in a distance of less than 2 miles from the level valley floor to a height of 7,500 feet. The highest peak of this range, Mount Grant, 10 miles northwest of Hawthorne, has an elevation of 11,303 feet, and Cory Peak, about the same distance southwest of the town, an elevation of 10,516 feet. The old Bodie stage road goes over the range about 2 miles south of Cory Peak through a pass whose summit is 8,000 feet above sea level. The Lucky Boy mines are located in rather low hills, which have an elevation of 6,225 feet, about one-half mile west of the place where the Bodie stage road begins the ascent from Walker Lake valley.

GEOLOGY.

Sedimentary rocks.—In the vicinity of Lucky Boy a series of metamorphosed light-colored thin-bedded, slightly cherty limestones, associated with calciferous sandstones and some dark shales, are exposed. Their structure is very much complicated by faulting and intrusion. In general these sediments seem to strike nearly east and to dip south at medium to high angles. These sedimentary rocks, interspersed in many places with intrusive rocks, extend south and southeast of the stage road for about 1½ miles. No fossils were found by the writer in these metamorphosed sediments, which are probably Mesozoic or younger in age, as it has been rather definitely shown that there are only small isolated areas of Paleozoic sediments between a nearly north-south line just west of Battle Mountain and the crest of the Sierra Nevada. This fact was first noted by the geologists of the Fortieth Parallel Survey.[1]

The Walker Lake valley is underlain by roughly stratified gravels and silts deposited during Pleistocene time in an arm of Lake Lahontan, as described by Russell.[2] Spurr[3] has noted rock-cut benches and gravels 400 feet above those described by Russell. He also noted gravels which he considers to be Pliocene in age on the Bodie road, at an elevation of 7,100 feet west of the summit.

Igneous rocks.—The limestones are intruded in a most intricate manner by a light to a medium gray granular rock, which is in turn cut by narrow dikes, some of which are dark colored, though many are light. The granular rock is composed of orthoclase, microcline, quartz, biotite, plagioclase, and hornblende, named in the order of decreasing abundance. It is as a rule quite fresh, though some secondary chlorite and sericite were noted in thin sections. Magnetite and apatite are accessory. The feldspars show as fairly large anhedral areas with smaller irregular areas of quartz between them. The plagioclase feldspar is between oligoclase and andesine, to judge from the extinction angles. This granodiorite is probably near granite in chemical composition.

The dark-colored basic dikes were seen in only a few places on the surface; the specimens are too much altered for determination, though they are seen to be basic phases of the granodiorite. The light-colored dikes are aplite, a fine-grained granular mica-free rock, composed of feldspar and quartz, and alaskite, a very coarsely crystalline feldspar and quartz.

Contact metamorphism.—The contacts between the granodiorite and limestone are sharp, and there has been considerable contact metamorphism along them, though this process has not formed distinctly marked zones. The limestones near the contact are altered to granular aggregates consisting of white garnet (grossular), tremolite, diopside, quartz, calcite, and a white chloritic material. The

[1] King, Clarence, Systematic geology: U. S. Geol. Expl. 40th Par., vol. 1, pp. 266–267, 1878.

[2] Russell, I. C., Geological history of Lake Lahontan: U. S. Geol. Survey Mon. 11. 1885.

[3] Spurr, J. E., Descriptive geology of Nevada south of the fortieth parallel: U. S. Geol. Survey Bull. 208, p. 117, 1903.

more or less parallel orientation of the minerals gives a slightly schistose appearance to some of the rocks immediately at the contact. Some distance north of the main contact there are small masses of epidotized limestone. As a rule the sandy sediments appear to have escaped much alteration, though in one slide of a calcareous sandstone the matrix shows some change.

ORE DEPOSITS.

LUCKY BOY VEIN.

History and ownership.—The veins at Lucky Boy were discovered in 1906 by men engaged in repairing the Hawthorne-Bodie stage road. The main claims, the Mountain King, Lucky Boy, and several others, were early acquired by the Goldfield-Alamo Mining Co., controlled by Messrs. Adams and Miller, merchants of Hawthorne. The greater part of the development has been done by lessees. The deep shaft, the Mountain King, in 1912 was leased to Mr. Charles E. Knox, of Tonopah.

Development.—The Lucky Boy vein is developed by several shafts, the deepest of which is on the Mountain King claim at the west end of the productive part of the vein. In July, 1912, this shaft was over 900 feet deep and sinking was under way to the 1,000-foot level. There are drifts both east and west of the vein at 100-foot intervals and a large part of the ore from the main shoot above the 900-foot level has been stoped. The 100 and 200 foot levels connect with the surface through eastward drift tunnels. The shaft is an incline on the vein to the 100-foot level, below which it lies 2 to 4 feet above the vein in the hanging wall. It is equipped with a 1-ton skip and all hoisting is done by electric power, generated near Bodie, Cal.

On the Lucky Boy claim, east of the Mountain King, there are several smaller shafts, the deepest of which is on the Haller lease. It is an incline on the vein and reaches a depth of 400 feet. A winze in the west drift on the lowest level was 80 feet deep. In 1913 a long crosscut had been driven from the Walker Lake valley to undercut all of the old workings.

Occurrence and general features of the vein.—The vein occurs in a rather irregular fracture, which strikes N. 80°–85° E. and dips 65°–75° S. This fracture lies near the north contact of a body of intrusive granodiorite in limestone, which shows the effects of contact metamorphism. The fracture took place after the intrusion and cuts indiscriminately the limestone, the granite, and the contact, though in general it has followed the contact rather closely. As a rule the hanging wall of the vein is granite and the footwall metamorphosed limestone, though there are many exceptions to this generalization. It was not a simple fracture. There are numerous splits in the vein, both in the hanging and foot walls, though the majority of the branches come in from the footwall side. In places there are lenses of somewhat mineralized, metamorphosed country rock between branches which reunite along both the strike and dip of the vein. The width of the vein ranges from 4 inches to 8 feet. At some places a difference of a few inches to several feet occurs in the space of 10 feet along the strike. The average width, however, is between 2 and 3½ feet.

Postmineral movement has produced gouge along the hanging wall, has crushed the vein material, and has formed a gouge on the footwall in a few places. This movement appears to have been nearly horizontal, though striæ which pitch to the east at low angles are seen here and there on the walls.

Occurrence and character of the ore.—The ore is not evenly distributed throughout the vein but occurs in small lenses and large shoots. These bodies have a persistent pitch to the west at rather steep angles and seem to be closely related to branch veins. The character of the wall rock apparently has little influence on the localization of the shoots. In the Mountain King ground the main ore shoot, about 120 feet long, was found at a depth of about 300 feet, though there were a few small lenses of ore in the upper workings. From the third level to a little below the fifth level the ore in this shoot was high grade. Between the sixth and eighth levels the vein was not heavily mineralized, though some small lenses were stoped. A short distance above the 900-foot level a second ore shoot has been opened, apparently at the junction of a branch vein with the main vein about 15 feet west of the shaft. The shoots and lenses do not end abruptly but pinch out to narrow streaks of good ore, which finally give out. The richer ore in the shoots occurs as bands, which are more often seen near the walls, particularly the hanging wall, than in the center of the vein. These bands range from 1 inch to 10 inches in width, but are between 2 and 4 inches at most places.

The low-grade ore, not mined at present, consists of crushed silicified wall rock, carrying more or less widely disseminated galena, tetrahedrite (gray copper), sphalerite, and pyrite. At the 900-foot level a lens of metamorphosed limestone between two branch veins contains a small quantity of the same minerals.

The medium-grade ore, which carries 50 to 400 ounces of silver to the ton, consists of fine-grained galena and tetrahedrite, together with a little pyrite. It occurs in bands up to 18 inches in width and in quite large bodies. The galena and tetrahedrite are finely granular, but in the richer ore they have been crushed, forming the "steel galena" and small fragments of tetrahedrite. As a rule the tetrahedrite is the common dark-gray to nearly black mineral, but in some stopes it has a light steel-gray color. The lighter-colored "gray copper" apparently contains a large proportion of silver in comparison with the copper content. The most valuable ore, which carries 2,000 to 3,000 ounces of silver to the ton, consists of light-colored tetrahedrite in a matrix of quartz and barite and occurs in veinlets in the ore shoots. In this grade of ore some of the tetrahedrite is coated with a thin black film which seems to be silver sulphide (argentite), and copper carbonate stains are also noted. One veinlet of this sort of ore on the 900-foot level shows also a little native silver. This is 300 feet below the ground-water level, which is at about the 600-foot level. Native silver has not been found above the 900-foot level, according to Mr. G. F. Badgett, superintendent of the mine.[1]

Enrichment of the ore bodies below the 300-foot level.—The Lucky Boy vein probably originally consisted of argentiferous galena and tetrahedrite irregularly distributed through a siliceous gangue. Postmineral movement fractured the ore and produced the well-marked gouge seen particularly on the hanging wall. Later the silver and copper content of the ores near the surface were taken into solution by atmospheric waters and carried down along the crushed veins to be redeposited by the tetrahedrite and galena in the vicinity of and below ground-water level. The theory of downward enrichment in silver veins has recently been added to by Messrs. Palmer and Bastin,[2] who have shown the precipitating effect of certain sulphides on silver carried in solutions such as are ordinarily found in mine waters.

Age of the vein.—The Lucky Boy vein cuts limestones of supposedly Mesozoic or younger age and granodiorite that has intruded these sediments. The granodiorite intrusion is probably to be correlated with the intrusion of similar rock which occurred on an extensive scale along the Sierra Nevada in late Cretaceous or early Tertiary time.

Milling.—The ore from the Mountain King vein is treated in a 10-stamp mill equipped with three Pinder concentrating tables, on which most of the lead is saved, and two Diester slime tables, which catch part of the gray copper that has a strong tendency to slime when crushed. Mine water is used in the mill, the available supply being about 55 gallons a minute.

OTHER PROPERTIES.

A number of less important lead-silver veins near the Lucky Boy were not studied in detail. Most of them are east-west fissures with southerly dip and have a mineralization similar to the main vein.

---

[1] Personal communication.

[2] Palmer, Chase, and Bastin, E. S., Metallic minerals as precipitants of silver and gold: Econ. Geology, vol. 8, pp. 140–170, 1913.

In the granodiorite mountains north of the Bodie stage road there are a number of prospects that could not be visited.

Cory Creek canyon.—Near the head of Cory Creek canyon, north of Cory Peak, there are said to be several eastward-striking veins in granite, none of which were being worked in 1912. Probably the "granite" is the granodiorite, the common rock on the west side of the Wassuk Range. Some of these veins carry lead and silver; others carry gold. There is said to be a 15-stamp concentrating mill at the Big Indian mine in Cory Creek.

Cat Canyon.—Near the forks of Cat Creek, about 7 miles north of west of Hawthorne, there is said to be about 1,500 feet of work on a copper deposit belonging to the Nevada Consolidated Mines & Selling Co. The deposit was discovered in the early eighties, and a little gold was extracted from surface ore. The country rock is granodiorite, cut by dikes of basic granodiorite and aplite. The ore occurs along an easterly striking zone of crushed rock that is reported to range from 25 to 100 feet in width, carrying from 1 to 4 feet of gouge on both walls. The material of this zone is somewhat siliceous and contains more or less disseminated pyrite and chalcopyrite; it is said to average 2½ per cent copper and $4 in gold to the ton. The development consists of three crosscut tunnels 60, 256, and 440 feet below the outcrop. In the upper crosscut a little chalcocite and bornite is reported. Between the 256 and 440 foot levels there is said to be a more or less open fracture next the hanging wall in which native copper is at present being deposited by the rather abundant water coming down along this zone. The vein is said to have been lost in the lowest tunnel at the west end of the drifts at a fault that has a northward strike.

La Panta and Pamlico mines.—South and east of Hawthorne there are many scattered mines and prospects, none of which were visited, and the few notes given below are gathered from various sources.

The La Panta and Pamlico mines, 10 miles east-southeast of Hawthorne, not visited by the writer, were actively worked in the eighties, and since that time considerable gold is said to have been recovered by various lessees. In 1912 a small amount of work was being done at each mine. Both properties belong to the Esmeralda Consolidated Mining Co., of Buffalo, N. Y. According to Messrs. Stannard and House, of Hawthorne, both veins occur in rhyolite near limestone and have a northeast strike. The mountains south of the mines are made up of light gray-blue and white limestones for a considerable distance. The ores from the two veins are similar, consisting of red and yellow stained sugary quartz, which carries rather coarse free gold as nuggets and wires, and some argentiferous galena. The galena, together with cerusite, is said to be found near bodies of high-grade free-milling gold ore. The gold is said to occur in relatively small bodies irregularly scattered through the veins.

The La Panta vein is said to range from a narrow stringer to 30 feet in width, averaging about 5 feet. It dips 35°–45° N. and is developed by a 300-foot vertical shaft and numerous drifts. The ore is said to occur in shoots and to be worth $15 to $17.50 a ton. It is said that this property has produced about $200,000.

The Pamlico vein, said to be nearly vertical, is developed by drifts, raises, shafts, and crosscuts aggregating between 4 and 5 miles in length. Its width is reported to be rather constant, but in some places it attains a maximum of 24 feet. The estimated gross production from the property is $500,000.

At the Pamlico there is a 20-stamp amalgamation mill of about 60 tons daily capacity. Water is piped from Cottonwood Creek, which heads near Buller Mountain in the Wassuk Range, 8 miles west of the mine.

Excelsior Mountain claims.—The Excelsior Mountain Copper Co. owns a group of claims 18 miles south of Hawthorne, at the south end of Whisky Flat. These claims were not visited by the writer, but the following notes, furnished by Mr. House, of Hawthorne, are of interest. The deposit was first worked in 1882 for copper ores, which were rich in silver and carried a little gold. The ores were smelted for a time in a 400-pound Mexican furnace on the ground. They occurred along a vertical eastward-striking contact of "granite" (probably granodiorite) intrusive into light-gray to white limestones that stand nearly vertical. The contact zone, said to be about 400 feet wide, shows a rough banding of constituents. Immediately at the contact there is said to be a narrow zone of very rich silver ore associated with chalcocite. For the 140 feet next north the metamorphosed limestones have little value. An 11-foot belt of copper carbonate ore associated with chalcocite and pyrite occurs next north, followed by about 80 feet of sulphide ore, pyrite and chalcopyrite, said to run between 3½ and 5½ per cent copper. Between the sulphide belt and the unmetamorphosed limestones there is a 200-foot belt of garnetiferous ore that is said to average 1½ per cent of copper a ton throughout, together with a little silver. This ore body is developed by several shafts and crosscuts.

SANTA FE DISTRICT, MINERAL COUNTY, NEV.

LOCATION AND ACCESSIBILITY.

The Santa Fe district (No. 16, Pl. I, p. 18), as the name is used in this report, comprises the part of the Pilot Mountains between New York Canyon and the Luning-Lodi road, and a small area in the northeastern part of the north arm of the Excelsior Mountains. All this territory is tributary to Luning, a town of about 75 inhabitants on the Hazen-Tonopah branch of the Southern Pacific Railroad. Luning is situated near the north end of the Soda Spring Valley, at the north end of a dry lake or playa, at an elevation of about 4,500 feet above sea level. This area lies in the east-central part of the region shown on the Hawthorne topographic sheet. The approximate distribution of the formations exposed in the district is shown in the upper right-hand corner of Plate XVIII, a sketch map of the geology of the Silver Star and Santa Fe mining districts. Good mountain roads radiate from Luning to most of the properties, though some of the mines in the Pilot Mountains have not as yet been reached by the roads. The ore taken from these places is sledded to loading platforms beside the roads. Water is very scarce in these mountains, and the few springs yield only a meager flow at best. At the properties in New York Canyon and in the Excelsior Range all the water is hauled from Luning. There are small springs near the Giroux, Santa Fe, and Fermina properties.

TOPOGRAPHY.

The Pilot Mountains form a bold escarpment on the east side of the Soda Spring Valley, which, in the area under discussion, trends about N. 20° W. (See Pl. XVIII.) The mountains rise to a height of 7,900 feet about 2 miles east of the valley, but 2 miles beyond they descend to about 6,500 feet in a chain of north-south intermountain valleys, such as those west of the Giroux mine (No. 6, Pl. XVIII) and between the Sunrise mine and Dumbarton spring (Nos. 18 and 3, Pl. XVIII). Spurr[1] considered that the scarp along the west front of the range is "very likely a simple fault scarp." Long, even-sloping débris cones issue from the larger canyons and extend nearly into the center of the valley. The fault scarp is greatly modified by the numerous deep canyons which cut it. Most of these canyons trend northeast near the base of the hills, but change to more nearly north after a mile or so. Their courses seem to be controlled in part by the structure of the range. The road running east to the Giroux mine, shown in the upper right-hand corner of Plate XVIII, skirts the southern boundary of the volcanic rocks, which appear to cover a large area north of the area shown on the map. In fact, Spurr[2] has said that the Gabbs Valley Range is largely composed of volcanic rocks.

---

[1] Spurr, J. E., Descriptive geology of Nevada south of the fortieth parallel: U. S. Geol. Survey Bull. 208, p. 103, 1903.

[2] Idem, p. 108.

GEOLOGY.

GENERAL FEATURES.

The rocks of the part of the Pilot Mountains visited by the writer consist in large part of crystalline limestones, which range from pure white to nearly black, and which are intruded by a granitoid rock between quartz monzonite and quartz diorite in composition. South of Volcano Peak there is exposed a series of red and dull brown argillaceous and calcareous conglomerates and sandstones which seem to overlie the limestone series. The low hills at the northeast end of the Excelsior Mountains southwest of Luning are composed of crystalline limestones, but the peaks are capped with andesite flows. The Gabbs Valley Range north of Soda Spring Valley appears to be made up entirely of lava flows, though the only portion of these rocks seen was along the road to Giroux mine.

SEDIMENTARY ROCKS.

PRE-TERTIARY (PROBABLY TRIASSIC AND JURASSIC) ROCKS.

General character.—The white and gray crystalline limestones of the north end of the Pilot Mountains yielded no fossils to the writer. In fact, in the part of the Santa Fe district covered by this reconnaissance it is very doubtful if organic remains will be found in the limestones on account of the highly crystalline character of the rock. The beds range from 2 to 12 feet in thickness. On the northeast flank of the Excelsior Mountains they strike about N. 45°–65° E. and dip 40°–70° N. In the Pilot Mountains, north of Volcano Peak, the prevailing strike of the limestones is N. 10°–20° W., but the dip ranges from nearly vertical to as low as 50° E. The structure is much complicated by extensive intrusions of quartz monzonite or quartz diorite, and further by numerous small fractures with either north or east strike.

At Volcano Peak, 4⅗ miles east of Luning, the limestones stand nearly vertical along a curved fault zone concave to the south. South of this fault the crystalline limestones appear to underlie a series of brown to red calcareous conglomerates, sandstones, and shales, which seem to extend for some distance south of the New York Canyon road. The relations are not entirely clear at this place, though it is reasonably certain that the red beds are younger than the limestones north of them, as the lower beds are conglomerates composed of bluish-gray limestone pebbles.

Age.—There is no direct evidence of the age of the crystalline limestones found in the Pilot Mountains east of Luning. They bear a strong lithologic resemblance, however, to limestones of supposed Triassic age at Lucky Boy and Yerington.[1] Spurr[2] describes "white

[1] Ransome, F. L., The Yerington copper district, Nev.: U. S. Geol. Survey Bull. 380, p. 102, 1908.

[2] Spurr, J. E., op. cit., pp. 99–101.

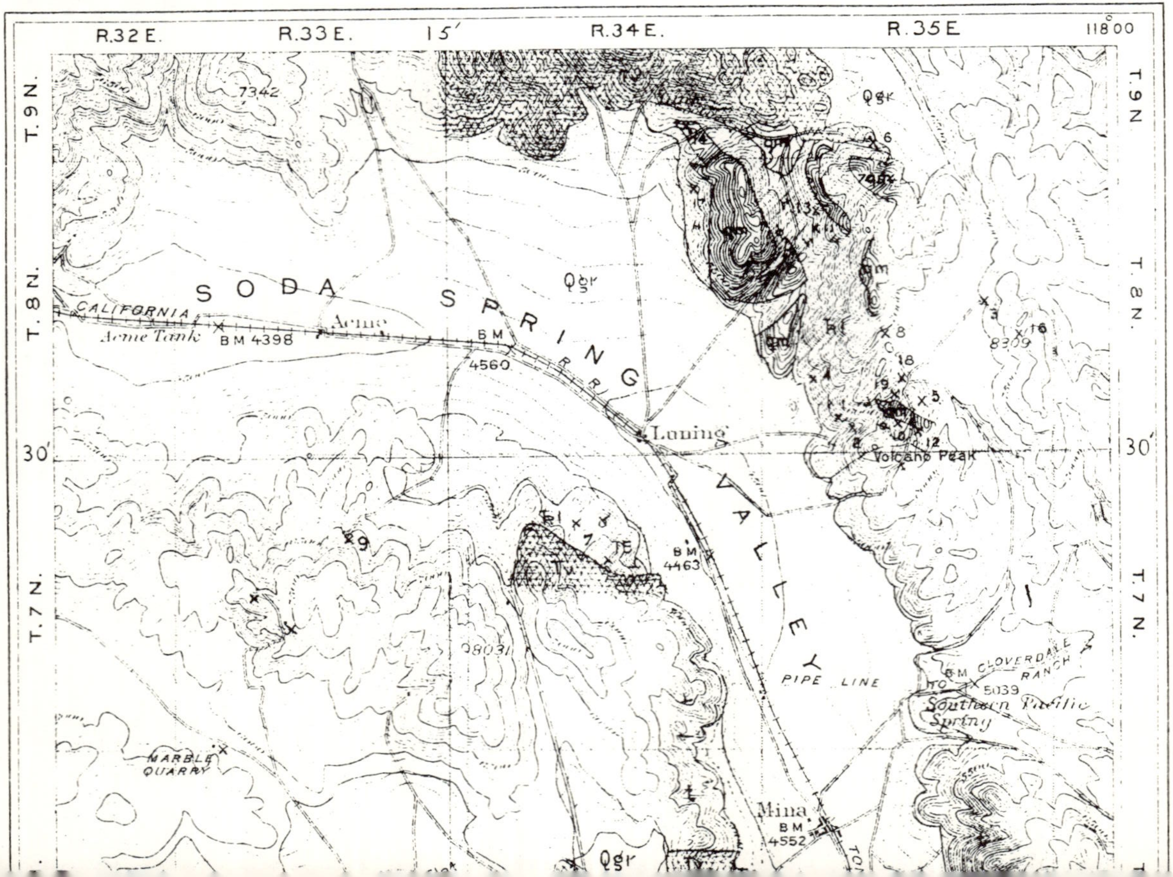

LEGEND

QUATERNARY

Qgr — Gravel and silt

Qb — Basalt

TERTIARY

Tgr — Pliocene gravel

Tv — Volcanic rocks

CRETACEOUS OR TERTIARY

qm — Quartz monzonite and allied rocks *(Intrusive)*

TRIASSIC? AND JURASSIC

t — Tuffs, argillite, and conglomerates

TRIASSIC

Ŧl — Limestone

Strike and dip

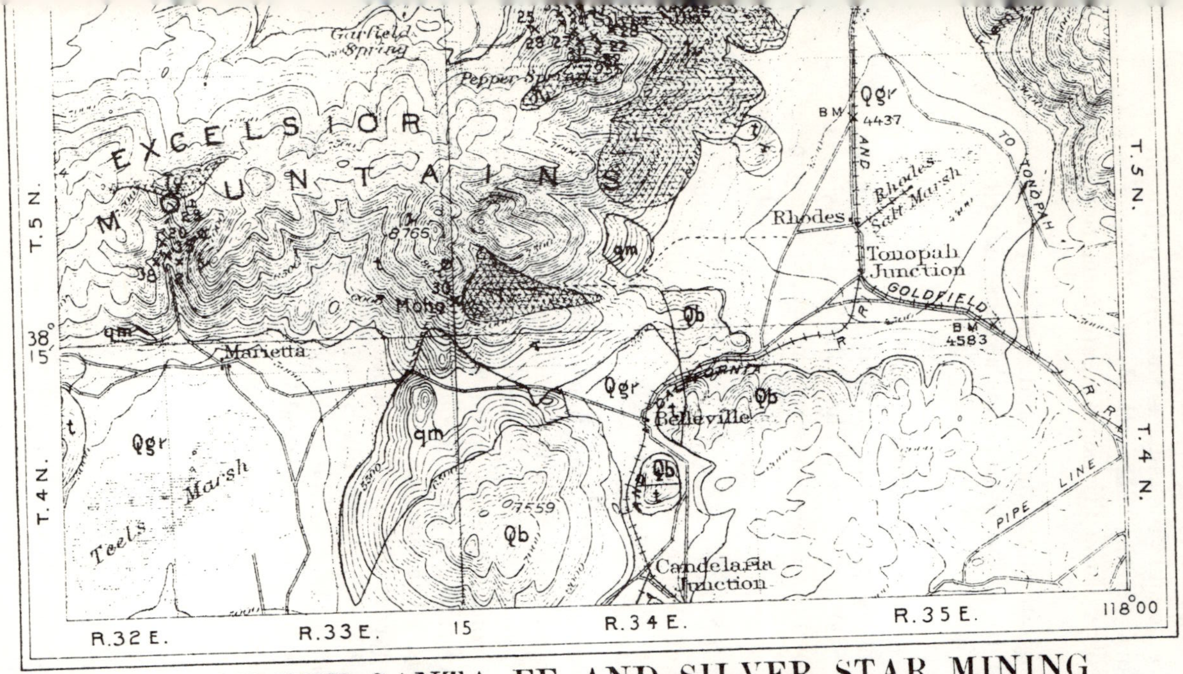

Mine and prospect

LIST OF
MINES AND PROSPECTS

SANTA FE MINING DISTRICT

1 CHAMPION (PARROT)
2 COPPER BUTTE
3 DUNBARTON
4 ESMERALDA
5 GARDIAN (F. M. SMITH)
6 GIROUX (ARIZONA-NEVADA)
7 INDEPENDENT
8 IRONSIDES
9 LOTTIE (DEL MONTE)
10 MAYFLOWER
11 NEVADA MARBLE CO.
12 NEW YORK
13 PILOT RANGE GROUP (FERMINA SARRIAS)
14 SANTA FE (HIDDEN TREASURE)
15 SHIPPER
16 SMOKED UP
17 TODD
18 SUNRISE
19 WALL STREET (TURK)

SILVER STAR MINING DISTRICT

20 BIRDSONG
21 BLACK HAWK
22 BOUNCE
23 ENDOWMENT
24 GEM
25 GRASSI
26 JUNIPER
27 HARDLUCK
28 LITTLE CHIEF
29 MARY
30 MOHO
31 NEW PARTY
32 ORIOLE
33 ORPHAN BOY
34 PEPPER
35 RIP VAN WINKLE
36 ROOSEVELT
37 SNOW BALL
38 WOODCHUCK

MAP OF THE SANTA FE AND SILVER STAR MINING DISTRICTS, MINERAL COUNTY, NEVADA

Showing location of mines and prospects and approximate distribution of formations

Base from topographic map of the Hawthorne quadrangle

Scale $\frac{1}{250,000}$

2 1 0 2 4 6 8 10 Miles

Contour interval 100 feet.
Datum is mean sea level.
1914

granular limestones" from Ellsworth Range, north of this area, and concludes that they are Triassic from their similarity to fossiliferous limestones discovered farther north in the same range by the geologists of the Fortieth Parallel Survey,[1] who describe the beds as "1,000–1,500 feet of dark grayish blue, compact earthy limestones of the Star Peak group," at the base of which "are dark-blue, finely laminated calcareous shales, rich in Triassic fossils." These limestones conformably overlie a series of green and purple quartzites and conglomerates, which are possibly the same as the series in the Silver Star district, west of Mina. In the Ellsworth Range this lower series is regarded as the equivalent of the Triassic Koipato formation of the Humboldt Range.

The red limestone conglomerates, shales, and sandstones along the New York Canyon road are probably the same as those described by Turner.[2] In limestones immediately overlying these beds he found Jurassic fossils. The writer found a few fragments of ammonites in red calcareous sandstones at an elevation of 6,400 feet on the New York Canyon road, about 1½ miles south of Mayflower Basin, concerning which T. W. Stanton says: "The collection consists of three fragments of ammonites, the best preserved of which seems to be referable to the genus Arnioceras, and hence indicates Lower Jurassic (Liassic) age."

Gabb[3] in 1869 described Jurassic ammonites from the Volcano District, 30 miles southeast of Walker Lake, which is probably the same locality as that in which the writer found the fossils identified by Dr. Stanton.

Alteration of the limestones.—The limestones wherever seen were crystalline, the sizes of grain ranging from nearly microscopic to as large as one-fourth inch in diameter. Along the various intrusive masses and dikes the contact-metamorphic zones are distinctly marked, but as a rule are rather narrow, though in some places they attain a width of 100 feet. The common alteration is that to a mass of light cream-colored garnet and some epidote containing quartz and more or less sulphide. In some places, as at the Mayflower mine, the limestone immediately at the contacts is altered to fine-grained light-green masses of microcrystalline quartz, tremolite, calcite, and some colorless garnet.

The outcrop of the Giroux ledge exposes a dark-gray brown-weathering jasperoid which contains light-colored angular fragments of cellular quartz.

---

[1] Descriptive geology: U. S. Geol. Expl. 40th Par., vol. 2, pp. 645–649, 1877.

[2] Turner, H. W., A sketch of the historical geology of Esmeralda County, Nev.: Am. Geologist, vol. 29, p. 267, 1902.

[3] Gabb, W. M., Am. Jour. Conchology, vol. 5, pp. 6–7, 1869.

QUATERNARY DEPOSITS.

The Quaternary is represented by a great accumulation of gravels, sand, and silt, filling Soda Spring Valley to an unknown depth. Large débris cones issue from the mouths of the different canyons and extend far toward the center of the valley. Their apparently even slopes are scarred by branching watercourses. Luning is built at the north end of a silt-covered flat or playa which, after the infrequent rains or in the spring, is covered with from 6 inches to 2 feet of water.

IGNEOUS ROCKS.

INTRUSIVE ROCKS.

General character.—The oldest igneous rocks in the Santa Fe district are quartz monzonites or quartz diorites, which occur as large masses and small dikes intrusive into the Triassic (?) limestones. None of these rocks were seen in the small part of the Excelsior Mountains included in this section, though it is said that at the Delmonte mine (No. 9, Pl. XVIII, p. 158) granular igneous rocks are intrusive into the limestones.

The larger masses of granitoid rock in the Pilot Mountains are composed of a gradation of types between quartz diorite and quartz monzonite, the central portions of the masses being in many localities the more basic. The smaller bodies are as a rule quartz monzonite porphyries, though some fine-grained aplitic dikes are probably end products of the same intrusive.

Petrography.—The large mass of granitoid rock running north from Giroux Canyon to the Hidden Treasure shows the widest variation of any of the bodies of the quartz monzonite in the district. Its edge phase is a light-gray, somewhat pinkish, granular rock, carrying only a small quantity of ferromagnesian minerals. The minerals, named in order of decreasing abundance, as seen under the microscope, are oligoclase-andesine, quartz, orthoclase, and green hornblende. Apatite is the most abundant accessory mineral, but titanite and magnetite are also present. The central part of the body shows a much finer grained basic phase of the intrusive, and is a quartz diorite that is locally somewhat porphyritic. In the porphyritic phases pink orthoclase and stout hornblende crystals are the phenocrystic minerals. The rock is composed of oligoclase-andesine, hornblende, quartz, and orthoclase. Magnetite is abundant, and apatite and titanite are less plentiful though common accessories. In this rock the ferromagnesian minerals exceed the quartz and orthoclase in abundance. The plagioclase feldspars of these rocks are everywhere zonally developed and occur as fairly well developed crystals, though few of them are complete. The other major constituents, as far as seen, never show crystal outlines. The quartz and orthoclase particularly occur in irregular grains, and the quartz forms an interstitial filling.

The smaller masses of the rock, particularly at the south end of the district in Mayflower Basin, commonly show a porphyritic tendency, though in many places the dikes are fine, granular, and light gray. The granular dikes are similar in mineral composition to the quartz monzonite phase of the intrusive, though the individual grains are small. In the porphyritic phases the phenocrysts comprise oligoclase-andesine, hornblende, and micrographic intergrowths of orthoclase and quartz. The groundmass of these rocks is composed of microgranular quartz and orthoclase and plagioclase feldspar, together with some small magnetite grains.

At the mouth of New York Canyon there is a sheet of light-gray much-altered latite porphyry. Whether this rock has any connection with the mineralogically similar quartz monzonite is not certain, though it may well have been derived from the same magma.

Alteration.—In general the igneous rocks are quite fresh, showing only slight surficial alteration, but near the ore bodies the quartz monzonite or quartz diorite as a rule is somewhat changed. Near the contact-metamorphic deposits silica has been added to the rock, and there has been a slight development of the heavy silicates, mostly epidote. Along the Hidden Treasure vein the basic phase of the quartz monzonite has been changed to some extent by the addition of brown biotite and the alteration of the feldspars to sericite.

VOLCANIC ROCKS.

The flow rocks north of the Giroux road, shown at the upper right-hand corner of Plate XVIII, are light-brownish to reddish-gray porphyries and tuffs showing flakes of biotite, blebs of smoky quartz, of which the largest are one-eighth inch in diameter, and white altered feldspars. These rocks are biotite rhyolite porphyries having brownish-stained glassy bases in which a few small quartz microlites are scattered. The brown biotite is somewhat altered to a reddish-colored uralitic mineral, but in one slide it seems to have been bleached to muscovite. Orthoclase is the most widely distributed feldspar, though some small scattered fragments of a calcic plagioclase are now largely altered to calcite, forming the white phenocrysts mentioned above.

The cap rock of the Excelsior Mountains, southwest of Luning, is a dense greenish fine-grained porphyry that weathers in ragged angular blocks of brown color. The groundmass is a dark glass thickly studded with minute fragments of plagioclase feldspar, magnetite, and what appears to have been particles of augite. The phenocrysts are much-altered, probably calcic plagioclase and augite that is altered to epidote and chlorite.

ORE DEPOSITS.

HISTORY AND PRODUCTION.

As far as can be learned the first ore bodies opened in the Santa Fe district were the De Monte (No. 9, Pl. XVIII, p. 158) and the Santa Fe, or, as it is now called, the Hidden Treasure (No. 14, Pl. XVIII).

In 1882 Burchard[1] mentions that the Farrington, probably the Fairmont, 8 miles west of Luning, was producing silver ores, and that at the Delaware, 6 miles east of Luning, copper ores were being mined.

The Santa Fe is said to have been discovered in 1879, and according to Burchard[2] was in operation in 1883, as was also the Sunrise mine (No. 18, Pl. XVIII, p. 158) and the Copper King, 6 miles north of Luning. He also notes that antimony occurred in the southwest face of Volcano Peak.[3]

During the year 1884 Burchard[4] says the Wall Street vein, 6 feet in width, was opened by a 150-foot tunnel and a 60-foot shaft, and that the 100-foot shaft on the Copper King had developed a 20-foot vein of $60 copper-silver ore. During that year the Sunrise ore was carrying about $125 in silver a ton.

Luning was for many years a station on the Wadsworth-Benton stage road before the advent of the Carson & Colorado narrow-gage railroad, which was the predecessor of the Southern Pacific. It is now a small settlement supporting a single store and having about 50 houses. The water supply is derived from a town well operated by a gasoline pump or in favorable weather by a windmill.

The silver veins were exploited for a number of years, but after 1893 little was done on those deposits. From 1906 to 1908 there was considerable development in the lead-copper ore deposits, but from 1909 to 1911 there was a period of stagnation. In 1912 the prospects of the district were much brighter, as a market for the copper carbonate ores was supplied by the blowing in of the Mason Valley Copper Co.'s smelter[5] at Thompson on Carson River. According to the reports in the mining journals the mines in the vicinity have shown much activity during 1913 and 1914.

---

[1] Burchard, H. C., Report of the Director of the Mint for 1882, p. 141.
[2] Idem, 1883, p. 513.
[3] Idem, 1883, p. 515.
[4] Idem, 1884, p. 349.
[5] Read, T. T., Mason Valley copper smelter: Min. and Sci. Press, vol. 105, p. 267, 1912.

No reliable figures of production from this district prior to the year 1906 are available; in fact there seems to be no publication in which any figures are given prior to those published by the United States Geological Survey in its annual volumes of Mineral Resources, from which the following table is compiled:

Production from the Santa Fe mining district, Mineral County, Nev., from 1906 to 1911, inclusive.

Year.	Ore treated.	Gold.	Silver.	Copper.	Lead.	Total value.
	Tons.		*Ounces.*	*Pounds.*	*Pounds.*	
1906	7,000	$34,837	2,900	....	....	$36,780
1907	9,489	41,824	8,056	105,199	64,000	71,579
1908	2,143	14,741	3,101	24,334	19,237	20,405
1909	409	5,403	19,088	34,877	20,000	20,724
1910	1,120	3,124	10,022	3,521	23,288	10,008
1911	158	1,191	2,902	1,256	26,393	4,074
	20,319	101,120	46,069	169,187	152,918	163,570

CHARACTER.

Typical contact-metamorphic deposits carrying ores of copper and lead are the most common type of deposits in the Santa Fe district. At some distance from any known intrusives there are a few bodies of replacement ores which are usually copper bearing, though the Julia of the Delmonte group carries lead. Veins are rather uncommon in the district, though the Sunrise deposit (No. 18, Pl. XVIII, p. 158) has the form of a vein cutting limestones, and there are a few small veinlike lodes in Mayflower Basin near the Guardian (No. 5, Pl. XVIII). The Hidden Treasure vein lies in altered quartz monzonite. It carried rich silver ores in its upper part but is low-grade pyrite below. The Nogal vein on the Todd group (No. 17, Pl. XVIII) is a lead-silver bearing vein in quartz monzonite.

All the ore deposits in this district, so far as known, are associated with the intrusive rock and are to be correlated with the late Cretaceous or early Tertiary period of mineralization enunciated by Lindgren.[1]

DEVELOPMENT.

There are few deep mines in the vicinity of Luning. Most of the properties are developed by open cuts, shallow shafts, and tunnels which do not attain great depths below the surface. The vertical shaft at the Giroux (No. 6, Pl. XVIII, p. 158) is said to be 700 feet deep and to have crosscuts at 200 and 600 feet, but it has been closed for at least a year and was not accessible. The flat incline at the Santa Fe is 300 feet long, but the face is scarcely more than 200

[1] Lindgren, Waldemar, Metallogenetic epochs: Econ. Geology, vol. 4, pp. 409–420, 1909.

feet below the surface. The vertical shaft at the Champion (No. 1, Pl. XVIII) is said to be 350 feet deep, and it is said that a crosscut at the 300-foot level opened up a 40-foot ore body that averaged 5.22 per cent of copper and 3 ounces of silver and 1 ounce of gold to the ton.[1]

MINES AND PROPERTIES.

CHAMPION GROUP.

The Champion group (No. 1, Pl. XVIII, p. 158) is situated at the mouth of a steep-walled canyon about 4½ miles due east of Luning. The mineralization has taken place along the south side of a dike of intrusive quartz monzonite that strikes about N. 40° W. The limestones immediately south of this mass have been altered. The ore zone, about 30 feet in width and 400 feet long, lies in greatly crushed altered limestone parallel to the dike. The ores, which are developed by numerous surface tunnels, open cuts, and pits, are all oxidized and consist largely of fragments of limestone more or less impregnated with malachite and chrysocolla. Azurite and cuprite are found in minute crevices in the ore, but no primary sulphides were noted. It is said that at the 200-foot level of the 300-foot shaft in the property the ore is still oxidized. In the ore zone there is not a large quantity of either garnetized or epidotized limestone.

COPPER BUTTE CLAIM.

The Copper Butte claim (No. 2, Pl. XVIII), just southeast of the Champion, is developed by open cuts and a short tunnel. The ore, consisting of oxidized copper minerals in slightly altered limestone, makes along a zone of fracture that strikes N. 40° W. and dips southwest near a small dike of biotite monzonite porphyry. During the summer of 1912 several shipments of 5 per cent ore were made from the property.

ESMERALDA.

The Esmeralda Copper Co.'s property (No. 4, Pl. XVIII, p. 158) was not visited. It is said that some good grade copper carbonate ore has been taken from replacement bodies on these claims.

GUARDIAN GROUND.

On the Guardian ground (No. 5, Pl. XVIII) and adjoining claims the limestones and quartz monzonite are both cut by narrow gash veins carrying quartz, chalcopyrite, and a little galena. They are

---

[1] McCormick, E., The copper deposits of southwestern Nevada: Min. and Sci. Press, vol. 81, p. 401, 1900.

undeveloped and the surface ores, though partly oxidized, still show the sulphides. In the limestone there are also a few replacement bodies of oxidized copper ores.

GIROUX GROUP.

The Giroux group of 33 claims (No. 6, Pl. XVIII) is controlled by the Arizona-Nevada Copper Co. The main development is situated on an eastward-trending 70-foot zone of dark siliceous brecciated lode material carrying copper carbonates that lies between quartz monzonite on the south and altered crystalline limestones on the north. The deposit is in part capped by later flows of biotite rhyolite. The 700-foot shaft could not be entered, but a southward crosscut 50 feet below the collar of the shaft shows the following section, beginning at the mouth: One hundred feet of rhyolite; vertical fault zone, 3 feet of crushed rock somewhat copper bearing; 70 feet of crushed iron-stained rock, much silicified and showing local pockets and stringers of copper carbonates; 80 feet of quartz monzonite softened and bleached for 20 feet south of the ledge material. The ore minerals are largely chrysocolla, malachite, and azurite, occurring as small veinlets and films between the fragments of the breccia. Here and there small pockets of cuprite are seen, and hematite is rather abundant. In the gulch north of the shaft garnetized and epidotized limestones are present, and contact-altered white and blue limestones were cut by the shaft, as shown by the material on the dump.

INDEPENDENT.

The Independent (No. 7, Pl. XVIII, p. 158) is about 3 miles southwest of Luning in low limestone hills. The sedimentary rocks dip to the northeast at medium angles and are cut by a fault zone that strikes N. 45° W. and dips 70° NE. In the fault zone there is about 4 feet of crushed, partly rounded, and cemented fault breccia. Next the hanging wall at a number of places there is from 8 inches to 1 foot of mineralized gougelike material that shows a few remnants of chalcopyrite. The main ore minerals are chrysocolla and malachite with some copper pitch ore. The ledge, developed by shallow shafts and tunnels which nowhere attain a greater depth than 50 feet, has been traced for 500 feet. Small shipments of ore in 1912 carried 5.2 per cent of copper and 3½ ounces of silver to the ton.

DEL MONTE PROPERTIES.

General features.—The Del Monte properties (No. 9, Pl. XVIII) were not visited by the writer. The following notes are taken from a report on the properties prepared by H. C. Cutler, of Reno, Nev., and furnished by Mr. G. I. Wright, of Luning: The mines are 6½ miles in an air line west-southwest of Luning, and 5 miles south of Seco, a siding on the Southern Pacific Railroad that is 5 miles west of Luning. A combination 25-ton amalgamation and cyanide mill, with "Lane mill," Wilfley tables, and Frue vanners, is located at Seco. Water for this mill is obtained by pumping from a 175-foot well.

The sedimentary rocks (limestone and shales) strike N. 40° E. and dip 60° N. Two dikes of intrusive rock (called diorite, but possibly a basic phase of the quartz monzonite) are found on the claims. A belt of shale 250 feet thick lies between the limestone belts in which the ore makes.

Lottie vein.—The southern vein, the Lottie, strikes about N. 50° E. and dips 58° S. It cuts through from limestone into shale and finally into the intrusive rock. Rich ore was stoped from the vein in the limestone, but it pinched out on entering the igneous rock and was entirely scattered in the shales. The vein is said to be rather narrow, ranging from 1 inch to 18 inches in width, the average being 1 inch to 2 inches. The valuable constituents, which are largely silver and some gold, are carried in a quartz gangue without sulphide minerals, as far as explored. It is developed by two tunnels to a depth of about 200 feet. The lower tunnel is about 900 feet long, of which 300 feet is crosscut.

Julia.—The Julia ores occur in the upper limestone, and, from the figures accompanying Mr. Cutler's report, they form irregular lens-like masses along a fissure that strikes N. 80° E. and cuts the formation. The surface ores were largely cerusite, but galena is found in the oxidized ore 40 feet below the surface. The galena is fine grained and is combined with some antimony. The ore zone is developed by a 700-foot tunnel and two shafts to a depth of 160 feet. The ore is said to carry between $8.50 and $20 in gold, silver, and lead, of which one-half is gold.

MAYFLOWER BASIN.

A small flat on the New York Canyon road, about 5½ miles due east of Luning, is known as Mayflower Basin. In this basin the crystalline limestones are much fractured, contorted, and intruded by numerous small dikes of quartz monzonite, besides the larger body shown on Plate XVIII (p. 158). At most places in the central part of the basin the limestones strike north and dip west at high angles, but at the south end, on the Mayflower and New York properties, the beds strike east and dip to the south.

Mayflower property.—There is a good deal of contact metamorphism of the limestones in this region, particularly on the Mayflower (No. 18, Pl. XVIII). At this place the whole top of a hill is altered to a greenish-white jasperoid containing a little garnet and epidote. Between the summit and the large mass of quartz monzonite shown on Plate XVIII there is a belt of cream-colored garnet, containing a little epidote and considerable copper carbonate and silicate minerals, which is at least 90 feet wide and 400 feet long. The property is developed by short tunnels and open cuts that attain a depth of about 50 feet, but are still in oxidized ore, though at one place a little residual chalcopyrite was noted. It is said that the main ore body, measuring 40 to 60 feet in width and 400 feet in length, has been thoroughly sampled and averages 5½ per cent of copper throughout.

New York mine.—The New York mine (No. 12, Pl. XVIII) is on the east side of the canyon. Irregular lenslike bodies of copper carbonate ores occur in a mass of garnetized and epidotized limestone on the south side of an eastward-striking dike of quartz monzonite. Shallow open cuts and tunnels constitute the development. A 50-foot winze from the lower tunnel about 75 feet below the outcrop is still in oxidized ores showing some chalcopyrite, pyrite, and magnetite in a matrix of epidotized and garnetized limestone.

Vacation.—The Vacation ground, southwest and west of the Guardian (No. 5, Pl. XVIII) shows three small areas of copper carbonate ores in garnetized limestone. One of these areas, about 3 feet wide, lies along a small dike of quartz monzonite porphyry that trends north and dips 80° E. The other two, each about 20 feet wide, are eastward-striking, nearly vertical zones of contact-altered limestones, though no intrusive can be seen on the surface in the immediate neighborhood of the deposits.

PILOT RANGE GROUP.

The Pilot Range group (No. 13, Pl. XVIII), more commonly called the Fermina mine, from the name of its owner, Fermina Sarrias, is situated on the west side of Giroux Canyon, 6 miles northeast of Luning. The contact zone, about 70 feet in width, runs N. 15° W. on the west side of a quartz monzonite intrusion. The limestones dip to the east at steep angles and are highly crystalline for 1,000 feet west of the contact, which dips about 75° E. Light cream-colored garnet is the most common contact-metamorphic mineral, but epidote, magnetite, and brown iron garnet are present in smaller amounts. Chalcopyrite and pyrite are intergrown with the contact-metamorphic minerals. The sulphides occur in lenses practically unaltered 15 feet below the surface, above which the ores are largely limonite, chrysocolla, and malachite associated with some unaltered sulphides. This contact zone is traceable for at least 1,000 feet, and at a number of places shows copper ores. The sulphide ore is said to carry 14 per cent of copper and $4 in gold and $3 in silver to the ton and the carbonate ore 15 per cent of copper, in carload lots as it is sorted for shipment to the Thompson smelter.

SANTA FE.

The old Santa Fe or Hidden Treasure mine (No. 14, Pl. XVIII) is situated in a box canyon at the north end of the Pilot Mountains, 6 miles north of Luning and about a mile east of Soda Spring Valley. The vein ranges from a fraction of an inch to 2 feet in width and is frozen to the walls. It strikes N. 60° W. and dips 30°–35° NE., parallel to a distinct sheeting in the quartz monzonite wall rock. The monzonite near the vein has been somewhat altered by the change of the hornblendes to brown biotite and the partial replacement of the feldspars by the same mineral. North of the road there is an incline on the vein 300 feet deep, with levels 75 and 100 feet below the collar. Drifts extend northwest at least 250 feet on both levels, but the southeast drifts under the canyon are quite short. In the upper level the ore is coarsely crystalline white quartz with drusy openings carrying pyrite, a large part of which is altered to limonite and some hematite. In places a little cerusite is present. At the lower level the ore is entirely coarse pyrite and quartz, with a few small masses of galena in the quartz removed from the iron sulphide.

There is a shaft on what appears to be a continuation of the vein on the south side of the canyon, but it had been abandoned and could not be entered.

SHIPPER COPPER CO.

The Shipper Copper Co.'s ground (No. 15, Pl. XVIII) is situated at the northeast end of the Excelsior Mountains, about 4 miles south-southwest of Luning. There are several short inclines on the ore zone that reach a maximum depth of 60 feet. The ore makes along a bedding plane in white crystalline limestone immediately below a bed of pink and white marble. The oxidized copper ores occur in irregular lenslike masses in a zone that ranges from 6 inches to 2 feet wide, dips 40° NW., and strikes N. 45° E. One-fourth of a mile south and east of the mine the limestone strikes east and dips to the north at medium angles.

LUNING SYNDICATE.

The property of the Luning Gold Mining Syndicate, locally called the Todd mines (No. 17, Pl. XVIII, p. 158) is situated on the west slope of the mountains, 5 miles northeast of Luning.

Nogal vein.—The Nogal vein, the northern vein of this property, on which much development work has been done, strikes N. 60° E. and dips 70° SE. in quartz monzonite. It is developed by a 150-foot shaft, connecting with a lateral from a 550-foot crosscut which runs in south of the vein but is nearly parallel to it. The vein is exposed by two laterals, the shaft connecting with one 500 feet from the mouth. In the lateral nearer the mouth there is a winze said to be 100 feet deep.

The vein ranges from 4 inches to 2 feet in width and has a thin parting on the hanging-wall side but is frozen to the footwall. The ore, largely iron and copper stained quartz carrying cerusite, galena, and a little gray copper, is said to carry $20 a ton in lead, silver, and gold.

Red vein.—The Red vein, one-fourth of a mile south of the Nogal, is an irregular replacement in crystalline limestones that dip east at medium to low angles. The lenses of ore occur in a zone of fracture that strikes N. 40° W. and dips 50° SW. The ore is a very heavily iron-stained cellular quartz containing some large pockets of soft brilliant-red iron oxide powder, said to carry between $10 and $40 a ton in gold and silver. It is developed by a 100-foot inclined shaft, which was bulkheaded at the 40-foot level in 1912 as stoping on a fair-sized ore body was in progress at that time.

SUNRISE MINE.

The Sunrise mine (No. 18, Pl. XVIII) is in a northwestward-trending vertical fracture in limestone, which is developed by an irregular shaft to a depth of 100 feet and by a long crosscut that cuts the shaft 50 feet below the collar. The ore occurred in a short shoot and is oxidized at the greatest depth attained. It consists of argentiferous galena and chalcopyrite largely altered to cerusite and copper carbonates. The surface ore is said to have carried very high silver values.

WALL STREET.

The Wall Street (No. 19, Pl. XVIII) is on the north side of a large mass of quartz monzonite about 100 feet north of the contact. The copper carbonate ores occur along an eastward-striking brecciated zone, 30 feet wide, in westward-dipping limestones. Copper pitch ore, malachite, chrysocolla, azurite, and cuprite, the principal ore minerals, occur in thin films and irregular masses in reddish jasperoidal limestone. The ledge is developed by two tunnels running east from Champion Gulch, the upper one 200 feet in length and the lower considerably longer, to judge from the dump. The lower tunnel is caved at the mouth, but some ore on the dump shows unaltered kernels of chalcopyrite in the oxidized ores.

OTHER RESOURCES.

It is said that stibnite and antimony oxides have been taken from small stringers and pockets on the slopes of Volcano Peak.

The Nevada Marble Co. operates two quarries near Luning, one about half a mile south of the Fermina mine on the east side of Giroux Canyon, 6 miles northeast of Luning. A white coarsely crystalline limestone seems to be the principal bed. It strikes northeast and stands vertical. To the southeast there is a bed of mottled white and dark-gray marble that is much finer grained. The beds are opened by cuts to a depth of about 20 feet. The marble is somewhat jointed, but some fairly large blocks can be quarried. This company also owns another quarry 2 miles southwest of Luning, in the foothills, and much nearer transportation.

SILVER STAR DISTRICT, MINERAL COUNTY, NEV.

LOCATION AND ACCESSIBILITY.

The Silver Star district (No. 17, Pl. I, p. 18) is in the south-central part of Mineral County, Nev., the position of its center being in approximately 118° 15′ east longitude and 38° 20′ north latitude. It embraces the southern or main part of the Excelsior Mountains, which lie west of the Soda Spring Valley and south of Garfield Flat, and is shown in the lower left-hand part of Plate XVIII.

The district is tributary to the towns of Mina and Sodaville on the branch of the Southern Pacific connecting Goldfield with the main line at Hazen. Mina is a division point on this railroad and is the larger of the two towns.

There are five camps in the district, three on the north side of the mountains, Silver Star, Grassie, and Roosevelt; and two on the south side, Moho and Marietta. Silver Star is the largest of these camps, though in July, 1912, the population numbered about 15, and some of the houses of the town were not occupied. The other camps are virtually "one-man" affairs, though at one time Marietta, while a station on the Carson-Bishop stage route, was a flourishing settlement. The ores mined near Marietta are shipped from Belleville, a siding on the narrow-gage branch of the Southern Pacific Railroad that runs between Mina, Nev., and Keeler, Cal.

The roads in the district are good and wagon transportation to the shipping places is not as serious a problem as it is in many camps.

WATER SUPPLY AND VEGETATION.

Water is scarce and the supply obtainable is not very good. At Mina there are wells, and at Sodaville a number of hot, highly mineralized springs, though all the drinking water for that town is piped from

Martin Springs, about 3 miles northeast in the Pilot Mountains. At Silver Star the water supply is obtained from small seeps south of the town and at a higher elevation. These seeps have been developed by shafts or tunnels to give a small though continuous flow. Grassie camp is supplied by a small spring, barely sufficient for domestic purposes. About three-fourths of a mile east of Moho, on the road to Rhodes Marsh, a spring has been developed to supply a meager flow of water. Several shallow wells at Marietta give an abundance of highly mineralized water.

The meager and stunted vegetation is typical of the desert country, though on the higher parts of the mountains there is a small stand of juniper and here and there a nut pine to relieve the general monotony. The lower hills support only very small sage and greasewood shrubs. After the rains the lowlands are covered with forage and are used to some extent for early spring grazing.

TOPOGRAPHY.

The Excelsior Mountains form an irregular group extending east and west, contrary to the usual direction of ranges in the Great Basin province. They are composed of two nearly parallel ridges partially separated by low flats but connected by ridges that lie north and south. The mines of the Silver Star district, as the name is used in this report, are all situated on the main south ridge, which is rugged and cut by deep narrow canyons. Garfield and Excelsior flats, north of this ridge, are about 5,500 feet, and Teels Marsh, to the south, is about 5,000 feet above sea level. (See Pl. XVIII, p. 158.) The Soda Spring Valley, east of the range, has a minimum elevation of 4,400 feet at Rhodes Marsh. The mountains rise abruptly west of this valley, probably along a fault that trends north and south.

The highest peak in the vicinity of the mines has an elevation of 8,766 feet and is about 3 miles north of Moho and 5 miles southwest of Silver Star.

GEOLOGY.

The main body of the Excelsior Mountains is composed of dark-gray bedded rocks ranging from fine-grained argillites to coarse conglomerates. This series is capped by volcanic material along the north and east sides and has been intruded by granitic rocks in two places on the south side of the range.

SEDIMENTARY ROCKS.

Triassic (?) rocks.—The bedded deposits, well exposed at Silver Star and in Endowment Canyon, are composed of a very hard, siliceous compact material, and in the field were supposed to be conglomerates and argillites of normal sedimentary origin. The argillites, as a rule, are fine grained and black to greenish-gray or purple. In a few of the coarser beds fragments of quartz and in some places feldspar were noted. Under the microscope, however, the rock is seen to be composed of particles of plagioclase, orthoclase, and quartz, with a little hornblende and biotite in some specimens. The general fine-grained type is a highly consolidated andesitic tuff or arkose.

A coarser phase of the same rock was thought in the field to be an intrusive dike rock of andesitic character. Thin sections, however, show that the material is all fragmental and that these rocks are medium fine-grained andesitic tuffs or agglomerates.

Interbedded with the fine-grained tuffs are thick beds of conglomerate, ranging from red to purple or white, and containing subangular pebbles, the largest of which are 3 inches in diameter. The darker conglomerates are composed of subangular fragments of andesite or rhyolite, though the two varieties of rock are not seen in the same bed. These pebbles are firmly cemented and break across rather than out of the matrix when the rock is fractured. In some beds the fragments are elongated by pressure.

The lighter-colored conglomerates, found near the top of the series, are composed of fragments of reddish chert and white quartzite, cemented by quartz into a very hard rock.

As far as could be made out in this brief reconnaissance the finer-grained red, green, and gray shaly tuffs are the lowest members of the series and are about 400 feet in thickness. Above these beds lies between 2,000 and 2,500 feet of medium-grained tuffs in 2 to 4 foot beds, interstratified with andesitic and rhyolitic conglomerates that in places attain a thickness of 50 feet. The uppermost beds are the massive cherty and quartzitic conglomerates, interstratified with many layers of light and dark quartzites. These beds are at least 800 feet in thickness.

The sedimentary series in the Excelsior Mountains is at least 3,500 feet thick, though its total thickness was not measurable in the short time taken for this reconnaissance, on account of the complicated faulting and folding which the beds have suffered.

The age of these sediments is not determinable in the Silver Star district. Spurr[1] considered the rocks to be early Tertiary in age, possibly the equivalent of the Esmeralda formation, but he states that there is considerable question as to their correlation on account of the extent of the induration and altered condition of the series, and that they may be either Mesozoic or Tertiary.

[1] Spurr, J. E., Descriptive geology of Nevada south of the fortieth parallel: U. S. Geol. Survey Bull. 208, pp. 110–111, 1903.

In the Augusta Range the geologists of the Fortieth Parallel Survey[1] found quartzites which closely resemble the Triassic Koipato formation, and dark-colored limestones, argillites, and greenish cherts, with Jurassic fossils in the limestone members.

The central part of the Desatoya Range they found to be composed of not less than 6,000 feet of greenish and purple cherty conglomerates, capped with about 1,000 feet of quartzites and conglomerates passing into slates, which they considered to represent the Koipato group. Green porphyroidal conglomerates are prominent features. These beds are overlain by dark compact limestones with a yellowish shaly layer at the bottom rich in Triassic fossils.

It seems probable from the descriptions of the Triassic cited above that the dark-gray, green, and purple tuffs and conglomerates of the Excelsior Range are of Triassic age and are to be correlated with the Koipato formation of the Humboldt Range.

Tertiary (Pliocene) and Quaternary deposits.—Spurr[2] noted the presence of horizontally stratified gravels and volcanic tuffs of Pliocene age in the low hills on the west side of Soda Spring Valley that lie unconformably on the upturned Triassic (?) sediments. These sediments are in large part covered by the later detrital wash which fills the whole of the valley. Beautiful examples of the Quaternary outwash cones are seen at the mouths of the large canyons.

INTRUSIVE ROCKS.

Quartz monzonite.—Intrusions of two ages were noted in the Silver Star district. The oldest of these intrusions is probably of Cretaceous age. Granular igneous rocks were seen at only two localities on the south side of the range in the vicinity of Moho (see Pl. XVIII, p. 158), though possibly there are other masses of this rock in the mountains. The rock is a light-gray, coarsely granular to porphyritic quartz monzonite, consisting of oligoclase, orthoclase, microperthite, quartz, brown biotite, and green hornblende named in the order of decreasing abundance, and a small amount of accessory magnetite and apatite.

Augite andesite.—The younger intrusive rocks were seen only on the north and east sides of the mountains north of Silver Star and in the localities where they form a part of the west wall of Soda Spring Valley. (See Pl. XVIII.) They are typical augite andesite porphyries of very fine grain and few phenocrysts, but most exposures are weathered a deep red-brown or yellow. The hills composed of these rocks are quite distinct, on account of their brilliant coloring, from those formed of the bedded tuffs and conglomerates. Most of

[1] King, Clarence, Systematic geology: U. S. Geol. Expl. 40th Par., vol. 1, pp. 281–284, 1878; Emmons, S. F., Descriptive geology, idem, vol. 2, pp. 649–659, 1877.

[2] Op. cit., p. 111.

the thin sections examined are very highly altered, but the original constituents seem to have been andesine, augite, some biotite, and magnetite. The feldspar is altered to sericite and calcite, and the femic minerals to chlorites and some epidote. The magnetite in many slides is changed to red-iron oxide.

EXTRUSIVE ROCKS.

Biotite andesite.—The top of the ridge south of Silver Star and at the head of Endowment Canyon is capped by a light-gray pink-weathering vesicular biotite andesite that shows distinct phenocrysts of plagioclase feldspar, biotite, and a few hornblende needles. The feldspar is andesine-labradorite, as shown by thin sections, and the biotite is greatly in excess of the hornblende. The groundmass of this rock, a grayish glass with a few feldspar microlites, greatly exceeds in quantity the few scattered phenocrysts.

Rhyolite.—On the eastern road from Silver Star to Mina, about 1¼ miles northeast of the former town, there is a small area of gray platy rhyolite that shows long needles of hornblende in the hand specimens. In thin section the microscope shows the groundmass to be largely composed of glass carrying orthoclase microlites in which are set a few small crystals of orthoclase, and the hornblende needles are seen to be largely altered to iron oxide.

Basalt.—A dark-gray, nearly black vesicular basalt, apparently a rather recent flow, overlies all the above-described formations and forms the large rounded hill west of Belleville. This basalt covers much of the country southwest of the area shown in Plate XVIII.

STRUCTURE.

At Silver Star the beds strike about N. 66° E. and dip 75°–80° S. Near the crest of the range south of the town the dip decreases to 30°. On the south side of the range, northeast of Moho, the beds stand nearly vertical or dip steeply to the north and strike N. 70° E. Between Moho and Belleville the strike is about the same, but the dark argillitic beds in this locality dip to the south. In Endowment Canyon, a few miles north of Marietta, the beds again dip to the north at very high angles or are nearly vertical. At the head of the canyon the light-colored quartzite conglomerates dip to the north at low angles immediately under the capping of glassy andesite.

The variations in dip noted above might be explained by the following structure: The north side of the mountains near Silver Star is the southern limb of an anticline striking east-northeast whose summit has been eroded to the level of Garfield and Excelsior flats. At the crest of the first ridge south of Silver Star, where the beds dip at lower angles, there is possibly a shallow synclinal fold, along which the glassy biotite andesite is found. South of this locality, along the main crest of the mountains, there is a tightly compressed anticlinal fold that apparently strikes north-northeast, which accounts for the steep dips or vertical beds in the vicinity of Moho and the lower part of Endowment Canyon.

ORE DEPOSITS.

HISTORY AND PRODUCTION.

The Endowment mine, in the southwestern part of the Silver Star district, was the first property worked in this vicinity. It was discovered shortly after the mines at Aurora were opened and for a number of years was a steady producer of rich silver-lead ores. The production from this mine is supposed to be in the neighborhood of $1,500,000, though no authentic figures of production can be obtained. The original owners finally abandoned the Endowment after they had supposedly taken out most of the ore, and in 1903 it was relocated by Joseph Rutty and R. L. Mason, of Marietta.

In 1874 and 1875 F. M. Smith operated a borax plant on the south side of Teels Marsh, and this industry was continued for a number of years, being abandoned only after the discovery of richer deposits in Death Valley, Cal.

The veins near Moho are of comparatively recent discovery and little work has been done on any of them.

The Blue Light or Garfield copper deposits, located about 10 miles northwest of Silver Star, on the south side of the northern ridge of the Excelsior Mountains, were worked as early as 1882, when Burchard[1] reports a production of 128 tons of ingot copper from the properties. The total production from the Garfield mine was estimated by him at $100,000.

This property was not visited in the course of this reconnaissance, but from the descriptions of the ore body it would appear that it is an irregular eastward-trending replacement in limestone.

In the vicinity of Silver Star the first ore was discovered in 1893 on the Duke claim, on the divide west of town, by Thomas Pepper, E. Grassie, and D. J. Robb. These men located 16 full claims and three fractions, covering a large part of the area then thought to be mineral bearing. The Bounce vein was discovered in 1894, and a five-stamp amalgamation mill of the California type was built that year to treat the ores of the district. The original locations were acquired by the Douglas Mining & Milling Co. in 1904, and since that time not much ore has been produced from that property, though a large amount of development work has been done. Several of the other properties have been steady producers. The total production from these mines

[1] Burchard, H. C., Report of the Director of the Mint for 1883, p. 514.

previous to 1901 was estimated by Mr. Douglas[1] at $300,000, from ore averaging $25 a ton.

The latest discovery on the north side of the mountains is the Roosevelt, located about 2 miles east of the town of Silver Star.

Figures of production from the district previous to 1902 are not to be found. The following table is compiled from the Mineral Resources reports, published by the United States Geological Survey:

Production of the Silver Star district, Mineral County, Nev., from 1902 to 1911, inclusive.

	Crude ore.	Gold.	Silver.	Copper.	Lead.	Total value.
	Tons.		*Ounces.*	*Pounds.*	*Pounds.*	
1902	480	$5,958	148			$6,038
1903	205	7,403	383			7,589
1904	256	4,671	19,208			15,097
1905	698	14,850	3,356	56,376	14,106	26,334
1906	650	6,000	5,200	180,000	120,000	51,064
1907						
1908	234	2,232	1,345	591	64,453	5,753
1909	127	862	11,915	315	45,581	9,060
1910	250	4,301	7,669	536	15,427	9,191
1911	121	1,352	4,829	35	6,103	4,190
	3,021	47,629	54,053	237,853	265,670	134,316

TYPES OF VEINS.

Two distinct types of veins occur in the Silver Star district. On the north side of the mountains near Silver Star there are a large number of eastward-trending veins that consist of white sugary quartz, adularia, and a little calcite or siderite, carrying gold and silver as the only metallic constituents. This type is well represented by the Bounce, Snowball, Jupiter, Mary, and New Party veins. Closely associated with these veins is the Roosevelt, which carries free gold in crushed, hydrothermally altered, andesitic rock, and possibly the Moho vein is altered andesite tuff. On the south side of the mountains, in the vicinity of Marietta, the eastward-striking veins contain a large quantity of the base minerals (galena and sphalerite), but the surface ores were extremely rich in silver.

QUARTZ-ADULARIA VEINS.

OCCURRENCE AND CHARACTER.

In general the quartz-adularia gold veins on the north side of the mountains strike not far from east and dip to the south at steep angles usually between 60° and 80°. At Silver Star there are numerous veins closely spaced and all very similar, which range in

[1] Turner, H. W., The mines of Esmeralda County, Nev.: Min. and Sci. Press, vol. 82, p. 73, 1901.

strike from N. 60° E. to N. 75° W., but with one exception they dip to the south at rather steep angles. At a few places northward-trending vertical veins were seen, but most of the fracturing along north and south lines seems to have taken place after the mineralization of the veins and to have been of relatively slight extent.

The veins occur in the conglomeratic beds of the sedimentary series or along the contacts between these rocks and the finer-grained rocks more commonly than in the finer-grained rocks, though some of the veins are in the medium fine grained andesitic tuffs. The veins are usually clean-cut fissures ranging from a fraction of an inch to 4 feet wide, but locally, as at the Bounce, they expand to a width of 13 feet between walls. The hanging wall of most of the veins is marked by clay gouge, presumably formed by postmineral movement. In a number of the veins there is also gouge on the footwall side. In the walls, and particularly in the footwall, there are as a rule a large number of small feeders that run into the veins from all directions.

The vein filling is in general largely composed of sugary white quartz somewhat stained with iron or manganese that has a peculiar platy or rhombic structure and many drusy openings. Most of these openings are coated with small clear quartz needles. Locally a little calcite or siderite is seen, but it is quite scarce. In a small vein west of the "Water Canyon," nearly opposite the New Party tunnel, there is a short tunnel on a calcite vein that strikes N. 45° E. in which some of the calcite is replaced by quartz. In practically all of the veins there is more or less filling that consists of angular fragments of silicified wall rock surrounded by crusts of quartz and adularia.

Under the microscope thin sections of the sugary quartz vein material are seen to consist of an intergrowth of cryptocrystalline quartz and adularia in the usual rhombic forms. In some slides or in parts of a single slide one or the other of these minerals may be in excess, but in general the quartz greatly exceeds the feldspar in bulk. These two minerals are clearly pseudomorphic after calcite. No sulphides are seen in these veins, but very fine flakes of free gold can be panned from most of the ore.

DEVELOPMENT.

A detailed description of the development on the different veins is hardly necessary, for most of them have not been opened below a depth of 100 feet and many are shown only by surface cuts and shafts.

Work on the Bounce (No. 22, Pl. XVIII, p. 158) has reached a greater depth than that on any other of the veins. A 400-foot shaft, now caved below the 75-foot level, is said to have extensive drifts at the 175, 200, 300, 350, and 400 foot levels. The Juniper (No. 26, Pl. XVIII) vein was being worked in 1912 through a 350-foot vertical shaft. The old work on this vein included tunnels west of and above the collar of the shaft along the croppings of the vein. The Snowball and Mary veins (Nos. 37 and 29, Pl. XVIII) are opened on the surface by a series of shafts and open stopes to depths of 100 feet and are also cut by the Watson tunnel, a 450-foot crosscut, which runs N. 43° W. from the bottom of the canyon just west of Silver Star. The Oriole vein (No. 32, Pl. XVIII) is opened by a 377-foot inclined shaft with drifts at the 50, 150, 250, and 350 foot levels.

The Gem and Pepper (Nos. 24 and 34, Pl. XVIII) are located east of Silver Star. The Gem is opened by several inclines and open cuts on the surface and a southward-bearing crosscut tunnel 570 feet in length, which does not appear to have struck ore. On the surface the ore occurs in two intersecting faulted silicified zones. One of these zones strikes north and dips 75° W., and the other zone strikes east and dips 50°–70° N. At the Pepper property there is a 100-foot vertical shaft in a zone of very highly altered, pyritized and sericitized andesitic rock. Whether the rock was originally the andesite tuff or intrusive andesite is not entirely clear, though it is thought to have been intrusive into the water-laid series.

The Roosevelt deposit (No. 36, Pl. XVIII) occurs in a zone of intensely crushed and altered andesite. The ore body, about 4 feet wide, strikes N. 10° W. and stands vertical. It consists of yellowish to reddish crushed altered andesite containing a very little quartz. The gold occurs in fine flakes and the ore is rich. The body has not as yet been opened below a depth of 25 feet.

At Moho (No. 30, Pl. XVIII) there are two veins about half a mile apart which strike N. 30°–45° E. and dip southeast at medium angles. The ore consists of brecciated, bleached, and silicified fine-grained andesitic tuff, somewhat stained by iron. The veins are 2 to 4 feet wide and the gold is entirely free.

AGE OF THE VEINS.

The quartz-adularia gold veins and associated gold-bearing veins in hydrothermally altered rocks are thought to have been formed after the intrusion of the andesite. They are to be correlated with the veins of similar mineralization common in western Nevada that are of late Tertiary age.

BASE-METAL VEINS.

The veins in Endowment Canyon, north of Marietta, are distinctly different from those seen in the other parts of the Silver Star district. They are inclosed in quartzose rocks, either black quartzite or white quartzite conglomerates, and not far removed from a mass of intrusive quartz monzonite. (See Pl. XVIII, p. 158.)

Endowment.—Two veins are exposed in the Endowment tunnel (No. 23, Pl. XVIII). At the mouth of the tunnel there is a vein that ranges in strike from N. 45° W. to N. 80° E. and dips 45°–50° S. At 340 feet from the mouth of the tunnel this vein was left to the south of the drift, which continues in barren ground for about 100 feet, where it strikes a fracture that trends N. 85° E. and dips 72°–87° N. The veins occur in quartzite and quartzite conglomerates with a few interbedded yellowish argillite layers. The southward-dipping vein is well mineralized for 280 feet from the mouth of the tunnel and has been stoped to the surface for that distance and below the level to a depth of at least 100 feet. It ranges from 2 to 6 feet in width. At and above the tunnel level the ores appear to have consisted largely of sand carbonate (cerusite), smithsonite, and a little copper carbonate and are said to have carried 8 to 10 per cent of lead and 15 ounces of silver to the ton.

A winze, said to be 350 feet deep, that is about 150 feet in from the mouth of the tunnel could be explored for a depth of 100 feet. Eighty feet below the level there is a 2 to 6 inch streak of gray copper next the hanging wall of the vein with sand carbonate and some unaltered sulphides in the ore below.

Some ore on the dump, said to have come from the bottom of this winze, consists of fine-grained galena, brown sphalerite, and small specks of pyrite and chalcopyrite. It is somewhat altered at this depth and carries irregular bunches of gray copper and a black mineral that appears to be silver sulphide.

The ore occurs as a replacement lode in crushed quartzite along a fracture zone and has much the appearance of a vein. The hanging wall, which seems to have limited the mineralization, is a thin bed of yellowish, somewhat calcareous argillite.

The nearly vertical fracture at the back end of the tunnel is 3 feet wide and contains crushed wall rock with heavy gouge on both walls. The mineralization along this zone has been slight, but there are some brecciated fragments of lead carbonate ore in the vein filling.

Birdsong.—The Birdsong property (No. 20, Pl. XVIII) is situated on the east side of Endowment Canyon, about a mile south of the Endowment mine. The vein, developed by a 200-foot drift tunnel and a 50-foot shaft, strikes N. 60° W. and dips 80° SW., cutting the quartzite beds at an acute angle. At the tunnel level the vein ranges from 2 inches to 2 feet wide, averaging about a foot, and contains massive galena and quartz with some anglesite and cerusite, said to carry 80 ounces of silver, $5 in gold, and 30 per cent of lead to the ton.

Fifty feet below the tunnel level the winze runs into the top of a mass of soft white rock that contains abundant crystals of pyrite largely altered to brownish limonite. This rock seems to be a sericitized quartz monzonite, though no rock of this type shows on the surface. At a point 150 feet from the mouth of the tunnel the vein is cut off by a fault that strikes N. 28° E. and dips 50° E. The movement along this fault has displaced the east side of the vein at least 50 feet to the south.

Rip Van Winkle.—The Rip Van Winkle vein (No. 35, Pl. XVIII) strikes N. 50° W. and dips 60° SW. It is a 2 to 4 foot zone of crushed quartzite that carries a small amount of galena and cerusite.

Woodchuck.—The Woodchuck (No. 38, Pl. XVIII) is developed by a 300-foot tunnel running northwest into the west side of Endowment Canyon. The ore makes along a zone of crushed quartzite that strikes N. 70° W. and dips 75° N. Pyrite is developed in the breccia and galena, pyrite, sphalerite, and chalcopyrite are seen in crevices in the footwall adjacent to the fracture. Iron-bearing potassium alum is being deposited on the walls by the mine waters.

Black Hawk.—The workings of the Black Hawk (No. 21, Pl. XVIII) were caved, and the dump, which is in a box canyon, was washed away. The vein strikes N. 60° W. and stands vertical in vertical beds of quartzite. The ores were said to have been sand carbonates and galena carrying silver. This property is supposed to be situated on a continuation of the Birdsong vein.

Other prospects.—In the low hills just north of Marietta there are some prospects in altered andesitic tuffs which show cuprite and copper carbonate ores that occur in crushed, altered, and bleached rock.

COTTONWOOD DISTRICT, WASHOE COUNTY, NEV.

In the section on Nevada in "Mining districts of the western United States"[1] there are three districts shown in central Washoe County, Nev., along the line of the Western Pacific Railway, namely, Cottonwood (Round Hole, No. 157), Deep Hole (No. 158), and Sheephead (No. 160). During the reconnaissance on which this report is based it was found that there are no mines on the west side of the Smoke Creek desert in the vicinity of Sheephead and Round Hole. At Round Hole there is a ranch and a thermal spring, but the mountains to the west are composed of basaltic flows in which no metals have been found.

The Cottonwood (Round Hole) district is erroneously located on the map in Bulletin 507 (Pl. IX). The Cottonwood district is sit-

[1] Hill, J. M., Mining districts of the western United States: U. S. Geol. Survey Bull. 507, 1912.

uated at the north end of the Fox Mountains, south of the Western Pacific Railway tracks. (See No. 18, Pl. I, p. 18.)

GEOLOGY.

The Fox Mountains form a northward continuation of the Pyramid Lake Range and were called the Lake Range[1] by the geologists of the Fortieth Parallel Survey, who report that[2]—

> Granites and Archean rocks occupy a large area at the northern end of the Lake Range, and but for the occurrence of small outbreaks of basalt that protrude through the granite and skirt the flanks on both sides the entire upper portion of the mountains might be so referred. * * * South and east of Pah-rum Peak the granite falls away rapidly and is soon concealed beneath heavy beds of dark shale, which have been provisionally referred to the Jurassic age.

The granites are composed of orthoclase, plagioclase, quartz, biotite, and hornblende and are referable probably to the quartz monzonite of late Cretaceous age. At the extreme northern end of the range occur a number of low rounded hills of characteristic brownish-gray gneissic formation that consist of triclinic feldspar, orthoclase, quartz, and hornblende, with large numbers of apatite crystals. These beds were referred to the Archean by the Fortieth Parallel Survey, though it seems more probable in the light of more recent knowledge that they may belong with the Jurassic and Triassic slates and schists common to the Sierra Nevada and western Nevada.

The Western Pacific Railway skirts the eastern side of the Smoke Creek desert about a mile from the base of the range and swings east around the north end of the range to Gerlach, the new town and a division point of the road. As seen from the cars the main mass of the mountains appears to be composed of granitic rocks, but on the western side there are some highly colored areas that from a distance look more like altered volcanic rocks. At the north end of the range, where the railroad turns eastward, the low hills south of the tracks are composed of consolidated sands and gravels that are probably the remnants of bench gravels formed during Lake Lahontan time. At some places along the western front of the Fox Mountains there is a suggestion of rock-cut benches that line up fairly well with the gravels mentioned above.

The Fox Mountains, still called the Lake Range, are shown on the Granite Range topographic sheet of the United States Geological Survey published in 1894. They rise abruptly from the flat desert valleys on either side. West of the mountains is the absolutely bare Smoke Creek desert, and east is the smaller San Emidio desert, an arm of the Black Rock desert.

---

[1] King, Clarence, Atlas: U. S. Geol. Expl. 40th Par., sheet No. 5.

[2] Descriptive geology: U. S. Geol. Expl. 40th Par., vol. 2, pp. 814–815, 1877.

The highest peak in the range, "Pah-rum," which rises to a height of nearly 8,000 feet, is 27 miles southwest of Gerlach and 11 miles north of Pyramid Lake.

The Smoke Creek and Black Rock deserts were visited by Russell in his studies of Lake Lahontan, as they were occupied by a part of that Pleistocene lake. He found that pre-Pleistocene faulting occurred along the west side of both of these depressions, along the eastern base of the Fox Mountains and the Pine Forest Mountains, which separate the Smoke Creek and Honey Lake Valley, and also that post-Pleistocene movement occurred along the later fault.[1]

The thermal springs along the western and northern borders of Smoke Creek desert, he states, occur along the post-Pleistocene fault zone.[2]

MINES.

A fruitless effort was made to visit the few scattered mines in the Fox Mountains. The oldest mines are located in Cottonwood Canyon, 7 miles north of Pah-rum Peak on the east side of the mountains, and a few silver prospects are located near the head of Rodero Canyon, about 3 miles northeast of the peak. The only mine which was being worked in 1912 was the White Horse mine near the head of a westward-draining canyon about 2 miles northwest of Pah-rum.

Burchard[3] reported that in 1882 there were 100 locations in the Cottonwood district, which was 125 miles from the nearest mills and at least 100 miles from a railroad.

From the best available information to be obtained at Gerlach, Nev., and Doyle, Cal., it would seem that the prospects on the east side of the mountains are all quartz veins in the granitic rocks. The valuable metals are chiefly silver and lead, carried in galena and some antimonial mineral, possibly gray copper. Some ore from the Cottonwood Canyon mines seen at Gerlach consists largely of galena and quartz together with some small crystals that appeared to be stibnite. In general the ore is said to have been of low grade, but some small rich pockets were found. The mines in Cottonwood Canyon were worked in the seventies, and at one time a 5-stamp mill was built in the district, but as water is very scarce it was not successfully operated. The mines have been closed since about 1900. At Rodero Canyon there is said to be a 14-inch quartz vein in granite, which strikes northwest and southeast and carries silver in galena and quartz.

At the Wild Horse mine a 5-stamp mill treats ocherous quartz ores, the gold and silver content of which is recovered by amalgama-

[1] Russell, I. C., Geological history of Lake Lahontan: U. S. Geol. Survey Mon. 40, Pl. III, p. 28, and Pl. XLIV, p. 274, 1900.

[2] Idem, pp. 50, 280–281.

[3] Burchard, H. C., Report of the Director of the Mint for 1882, p. 166.

tion. The veins are said to strike northwest, to have a rather flat northeast dip, and to occur in softened altered porphyries. Whether the so-called "porphyry" is altered porphyritic quartz monzonite or was originally andesite is not clear. Some specimens of the rich ore from these veins show a highly silicified porphyritic rock containing plagioclase feldspars. The "vein" is said to average about 3 feet in width and consists largely of yellowish "altered porphyry" cut by little stringers of dark brownish-red iron-stained quartz. The yellowish ore is low grade, but pans very fine colors. The narrow siliceous streaks constitute the rich ore and pan flaky gold, some of the flakes being one-sixteenth of an inch across. The Wild Horse mine is owned by the Washoe-Lassen Mining Co., composed almost entirely of Susanville, Cal., people. The vein was located about 1902.

PEAVINE DISTRICT, WASHOE COUNTY, NEV.

LOCATION AND ACCESSIBILITY.

The Peavine district (No. 19, Pl. I, p. 18) covers an area about 16 miles from east to west and 8 miles from north to south, lying in the hills immediately north of Truckee River, at the extreme western side of Nevada. According to Browne,[1] as originally laid out in 1863 it was 20 miles from east to west by 10 miles from north to south, with a spring, at which a station was located, near the center. This station was probably Poeville, for that is the only town which was located in the area shown in Plate XIX.

At the west side of Lemmon Valley, shown in the upper left-hand corner of Plate XIX, there was at one time a stage station on the Beckwith Pass route from Reno to Sacramento. There are still a number of springs at this place, but the station is entirely demolished. This region is shown in the southwest corner of the Reno topographic sheet. Reno, the supply point for a large part of western Nevada, on the main line of the Union Pacific, is situated at the center of the south side of the district in Truckee Meadows. In late years the eastern end of the Peavine district has been known as the Wedekind, from the mine of that name located north-northwest of Sparks.

The mines in the northwest part of the district are served by the Nevada-California-Oregon narrow-gage railroad, the principal shipping point being Purdy, Cal., which is about 3 miles west of the area shown in Plate XIX.

TOPOGRAPHY.

Peavine Peak, with an elevation of 8,270 feet, is the highest point in the district. To the south the sides of this mountain rise in gentle

[1] Browne, J. R., Mineral resources of the States and Territories west of the Rocky Mountains for 1868, p. 316.

slopes from Truckee River, but on the northeast side there is an abrupt rise from Lemmon Valley. East of the Reno-Poeville road the hills are low and gently rolling, rising only a few hundred feet above Truckee Meadows, whose elevation is about 4,500 feet. A considerable area west and northwest of Reno is marked by flat-topped ridges.

GEOLOGY.

GENERAL CHARACTER OF THE ROCKS.

The oldest rocks in this area are schists, composed of highly compressed sediments and irruptive rocks. They form the major part of Peavine Peak. A rather coarse granular, medium to dark gray rock is intruded into the schists in an irregular manner. Many small dikes of this rock are seen in the schists but are in general too small to be shown on the map. A series of andesitic flows that are exposed in the low hills north of the Truckee Meadows overlie the granite and schistose rocks. Associated with the lavas but younger than most of the flows are partly consolidated sands and gravels that are seen west of Reno on both sides of the Truckee. Truckee Meadows and Lemmon Valley are filled with early Quaternary and more recent wash gravels, sands, and silts.

PRE-TERTIARY SCHISTS.

Previous descriptions.—The geologists of the Fortieth Parallel Survey state that[1]—

Peavine Mountain is formed of a series of highly altered quartzites and fine-grained feldspathic rocks, which have been referred to the Archean series. * * * They stand at a highly inclined angle with a strike varying from N. 50° to 65° E. * * * The entire series of beds is characterized by a fine-grained texture and all the beds seem more or less decomposed, the purer quartzites being penetrated by fissures and cracks filled by ferruginous material. In the quartzites are also minute grains of magnetite and occasionally a little yellowish-green epidote.

On the north slope of Peavine Mountain a much fresher rock carries, orthoclase, plagioclase, and both hornblende and mica.

Becker[2] describes a series of metamorphosed sediments in the vicinity of Steamboat Springs which he considers comparable with the Jurassic and Triassic rocks of the Fortieth Parallel Survey. This region was visited by the writer, and the formation, in external appearance, bears a close relation to the schists of Peavine Mountain.

In the area described in the Truckee folio, which is southwest of Verdi (Pl. XIX), Lindgren[3] has described a series of gray to black

---

[1] King, Clarence, U. S. Geol. Expl. 40th Par., vol. 1, p. 97, 1878; vol. 2, pp. 850–851, 1877.

[2] Becker, G. F., Geology of the quicksilver deposits of the Pacific slope: U. S. Geol. Survey Mon. 13, pp. 128, 129, 333, 334, 1888.

[3] Lindgren, Waldemar, U. S. Geol. Survey Geol. Atlas, Truckee folio (No. 39), 1897.

U. S. GEOLOGICAL SURVEY
GEORGE OTIS SMITH, DIRECTOR

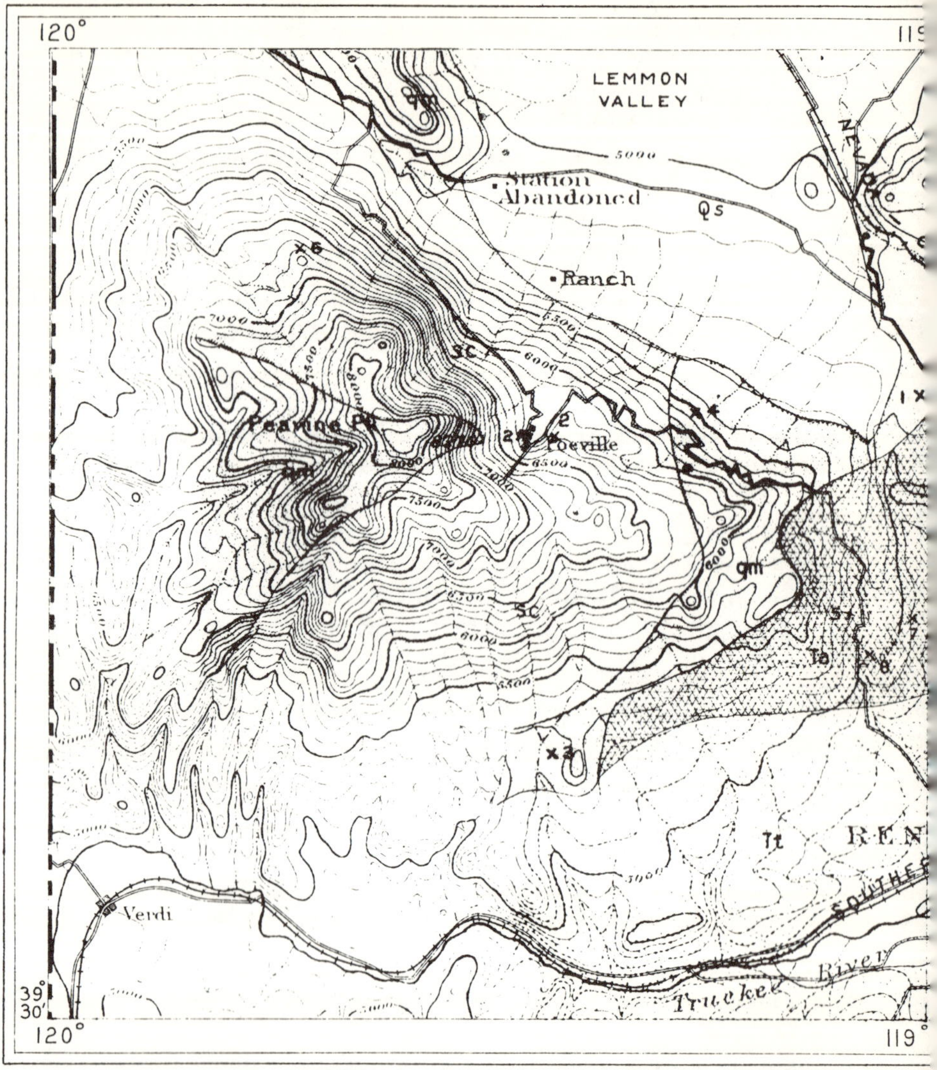

SKETCH MAP OF THE PEAVINE MINING DI

Base from topographic map o

Scale $\frac{1}{125,000}$

1 0 1 2

Contour interval 1(

Datum is mean sea

1911

QUATERNARY

Qs — Sand and silt

TERTIARY

Tt — Truckee formation *(Gravel, sand, and silt)*

Ta — Andesite flows

Quar

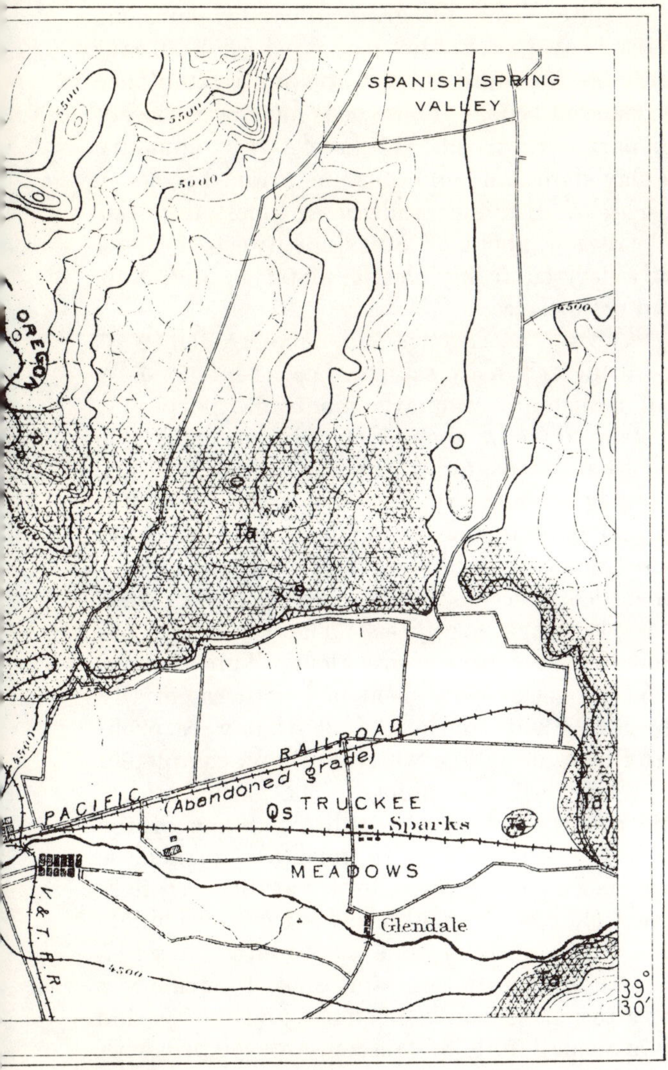

LIST OF MINES

1 EMMA
2 FRAVEL
3 NEVADA CENTRAL
4 NEVADA INDUSTRIAL PLACER
5 RECALL
6 RED METALS
7 RENO MIZPAH
8 UPDIKE
9 WIDEKIND
10 FULTON'S QUARRY

T, WASHOE COUNTY, NEVADA

quadrangle

4 5 Miles

S
Y (?)

PRE-TERTIARY

SC

Schists

×10

Mine or prospect

nite

siliceous sandstones, clay slates, and fine-grained banded siliceous rocks that weather a dark brown or reddish color, which he tentatively refers to the "Juratrias" because of the "striking resemblance of the rock to the Sailor Canyon beds." These beds have been closely folded and the folds in part overturned. The rocks are commonly micaceous schists, black clay slates, and siliceous fine-grained rocks.

Diller[1] describes a series of Jurassic sandstones, limestones, and slates in the Taylorsville region, about 65 miles northwest of Reno. The sandstones are partly derived from volcanic material, and show abundant plagioclase and orthoclase.

Distribution and character.—The eastern and northern flanks of the Peavine Mountain are underlain by schistose rocks intruded by numerous small dikes of monzonite or quartz monzonite porphyry. The schists have a persistent strike of about N. 60° E., and as a rule dip to the southeast at rather steep angles or stand nearly vertical. They have a characteristic green-gray to dark-gray color but weather a dull red-brown. Epidote is common in these rocks, particularly along the dikes. Some beds have been changed to muscovite schists, whereas others are siliceous schists, and still other beds are finely laminated, nearly black argillites. The rocks are not schistose in all places, but that structure is prevalent. Thin sections of some of the lighter-colored rocks show them to consist essentially of bands of quartz, alternating with darker bands, that were probably feldspars, and ferruginous minerals, which are now completely altered to chlorite and yellowish-green epidote. The darker schists are in general very fine grained siliceous argillites. Some of the beds of intermediate color are composed of fragments of orthoclase and plagioclase. These rocks show only slight parallel arrangement, and the feldspars are changed to kaolin and sericite. Epidote is developed to some extent, and secondary quartz produces the schistose banding. A further alteration of this kind of rock develops a mica schist, composed of alternating bands of quartz and feldspathic material in which there are small foils of muscovite developed with the long axis parallel to the schistosity.

INTRUSIVE IGNEOUS ROCKS.

A rather coarsely granular feldspathic rock is intrusive into the pre-Tertiary schists. The distribution of this rock, as shown on Plate XIX, is rather regular, but as a matter of fact numerous small porphyritic dikes of similar mineralogic composition are found in the area mapped as schists.

In the large areas shown on the map the rock has a dull-gray color and weathers in rounded drab-colored hills. Just east of Fulton's

---

[1] Diller, J. S., Geology of the Taylorsville region, Cal.: U. S. Geol. Survey Bull. 353, pp. 34–57, 1908.

quarry, 2 miles north of Reno, there is a small knob of this rock entirely surrounded by the volcanic flows of Tertiary age, which rest on the deeply weathered, eroded "granite."

The typical coarse phase of this formation shows plagioclase and orthoclase feldspars with rather abundant biotite and hornblende in hand specimens and in places a little quartz. Thin sections show the rock to be composed of oligoclase, orthoclase, quartz, brown biotite, and dark-green hornblende. The minerals show no crystal outlines except the accessory magnetite and apatite, though some of the ferruginous minerals show partly developed crystals. The feldspars are much more abundant than the other constituents, and the femic minerals exceed the quartz in relative amount.

In the dike rocks that cut the schists a decided porphyritic tendency is developed in the quartz monzonite and no quartz is seen except in the groundmass. In fact some dikes show practically no quartz, even under the microscope, and so pass into the monzonite porphyry class.

At the Nevada Central mine on the south side of Peavine Mountain the quartz monzonite has been altered to a soft, white mass of quartz, sericite, and pyrite, presumably by hot solutions.

TERTIARY VOLCANIC ROCKS.

Distribution and general character.—Andesites of Tertiary age overlie the eroded quartz monzonite on the east side of Peavine Mountain and from the low rolling hills north of Truckee Meadows. In general these rocks are dark basic lavas of fine grain, though as a rule they show a slight porphyritic texture. Along the old grade of the Southern Pacific Railroad, shown at the right-hand edge of Plate XIX, the andesites are vesicular and are easily mistaken for basalt. In the low hills 2 miles north of Reno, at Fulton's quarry, a coarsely porphyritic, light-gray andesitic flow overlies this andesite.

In this reconnaissance only these two types of andesitic flows were seen, though Mr. Jones,[1] of the University of Nevada, states that he has found three distinct flows, the oldest of which is a hornblende andesite in which the hornblende phenocrysts in places reach a maximum length of 2 inches. This rock is succeeded by the fine-grained augite andesite already mentioned, and this in turn is succeeded by the coarse porphyritic hornblende andesite of Fulton's quarry.

Augite andesite.—The augite andesite is the most widely distributed rock of the Tertiary volcanic area shown on Plate XIX. It weathers to a rusty reddish brown except where it has been hydrothermally altered. The fresh rock is a nearly black, fine-grained feldspathic rock with very small phenocrysts. Thin sections show

---

[1] Oral communication.

the groundmass to be composed of minute plagioclase laths, augite, magnetite, and titanite. The phenocrysts are largely andesine, though some crystals have extinction angles nearer labradorite. Augite in grains and some well-developed crystals was found as phenocrysts in the slides, and a few large magnetite grains were noted. The pale-green augite phenocrysts are somewhat altered to green hornblende, chlorite, and magnetite, but the feldspars are practically fresh.

The later flow, seen at Fulton's quarry (No. 10, Pl. XIX), is a light-gray, coarsely porphyritic rock, which in hand specimens shows phenocrysts of plagioclase, hornblende, and biotite. Many of the feldspars are a quarter of an inch long, the biotite flakes as large as one-eighth of an inch in diameter, and the hornblendes generally occur as small needles, though in some specimens they attain a maximum length of one-half inch. Under the microscope the groundmass is seen to consist of small grains and crystals of the same minerals with a considerable quantity of quartz. The presence of the quartz was not suspected in the field and the rock was considered to be an andesite porphyry, but as it contains such an appreciable amount of quartz it should be called a biotite-hornblende dacite.

Hydrothermal alteration of the andesites.—As far as seen, the dacite has not suffered hydrothermal alteration, but it may possibly have been so affected at some place in the district.

The augite andesite has suffered the most intense propylitic alteration in general along lines that trend east and west, though near the east base of Peavine Mountain the zones of alteration as a rule trend north and south. The alteration is of two overlapping kinds, as pointed out by Louderback.[1] In one type of alteration the rock is bleached and softened with the introduction of abundant pyrite. This results in zones of soft light green to brilliant red or white rock that is largely a mixture of sericite, calcite, and pyrite, or iron oxide. The other type is represented by the irregular elongated ribs of siliceous material trending east and west that are seen particularly in the hills east of the Nevada-California-Oregon railway tracks. These zones appear to have occurred along fractures which have brecciated the andesite, as the silica cements fragments of that rock which themselves have been in some places almost completely replaced by silica. It was at first thought by the writer that these reefs were dikes of rhyolite cutting the andesite, but every gradation between unaltered andesite and pure silica was found. In general the replacement has not been so complete as to obscure the porphyritic character of the andesite.

[1] Louderback, G. D., General geological features of the Truckee region, east of the Sierra Nevada: Geol. Soc. America Bull., vol. 18, pp. 664–665, 1907.

Along these zones of altered andesite, and particularly those of the first type, the ore deposits of the eastern end of the Peavine district are located.

TERTIARY SEDIMENTS.

Truckee formation.—Beds of partly consolidated gravel, sand, and silt are present in the Truckee Valley west of Reno as far as the California line. These beds are inclined on the south side of Peavine Mountain and dip toward the river at angles of 10°–15°. They underlie the flat-topped ridges which slope off from the main mountain mass. This series is at least 500 feet in thickness in the Truckee Valley, though its maximum thickness is not known. Seams of impure coal are reported in these beds north of Verdi.[1] Louderback[2] states that these gravels overlie the andesites of Tertiary age. They are regarded as of Miocene age.

QUATERNARY DEPOSITS.

Truckee Meadows and Lemmon Valley are underlain by fine sands and silts of Quaternary age. It is possible that in Lemmon Valley these late sediments may obscure older gravels and sands, though no deposits clearly referable to the Truckee formation were noted in this reconnaissance. In the Truckee Meadows the underground water is very near the surface and near the mouth of Truckee Canyon, east of Glendale (see Pl. XIX. p. 184), much of the land is marshy.

STRUCTURE.

The schists of Peavine Mountain have an approximate strike of N. 60° E., and in most places dip to the southeast at high angles, though on the northwest side of the mountain dips as low as 20° were recorded. They also appear to pitch southwest along the strike of the beds. The schistose structure is not everywhere well marked, but seems to be more pronounced near the large and small masses of intrusive quartz monzonite. It seems probable, therefore, that in large part the schistosity was developed by the intrusion of that magma. Following this intrusion there seems to have been a relatively long interval of erosion prior to the outpouring of the andesitic lavas, as along the contacts the granular rock is deeply weathered, even underneath the flows.

After the andesite extrusion, though it seems possible that the movement may have started before that time, faulting took place along the northeast side of Peavine Peak and formed the well-marked scarp overlooking Lemmon Valley. This fault strikes about N. 45°

---

[1] U. S. Geol. Expl. 40th Par., vol. 2, p. 849, 1877.

[2] Louderback, G. D., General geologic features of the Truckee region east of the Sierra Nevada: Geol. Soc. America Bull., vol. 18, pp. 665–666, 1907.

W., contrary to the general north and south course of the faulting in this vicinity, as pointed out by Louderback.[1] This fault zone goes through the pass just west of the abandoned station shown in the upper left-hand corner of Plate XIX. At the east side of the mountain it seems to have turned southward and split, as between the granite area and the Nevada-California-Oregon Railway tracks there are two zones of faulting that strike a few degrees west of north and are marked by very highly altered crushed belts of andesite.

The brecciated zones that stretch east and west in the low andesite hills north of Truckee Meadows were probably formed at or shortly after this major movement, which uplifted the Peavine Mountain block.

ORE DEPOSITS.

HISTORY AND PRODUCTION.

The earliest report of mining operations in the Peavine district was by Browne[2] in 1867. He states that gold and silver occur in quartz veins associated with copper carbonate ores. He reports that some copper, which carried $150 in gold and $250 in silver, was made at Reno by Dr. Landszwert from these ores. At that time the Bay State Mining Co. was intending to install a Haskell smelter at Reno.

Browne[3] reported in 1868 that there were 13 veins opened in the district and that they strike about north and south. cutting granite and metamorphosed sediments.

Hague[4] in 1870 says of the Peavine district:

> Veins of copper ore have been opened in metamorphic rocks and prospected to a depth of 40 or 50 feet. The surface ore consists chiefly of oxides and carbonates, with some sulphurets of copper. It is slightly argentiferous. The region has been lying neglected for several years, but its proximity to the railroad may make it an available source of copper should future exploration encourage further development.

Burchard[5] says that in 1881 the Golden Fleece at Poeville had opened an 8-foot vein by tunnel and incline. He also calls attention to the Massalond ledge 3½ miles west of Reno (possibly the Nevada Central), which is a 25-foot lode impregnated with pyrite that carries a little gold. He says that the copper belt of the district is 15 miles northwest of Reno, presumably on the northwest side of Peavine Peak.

Since 1908 there has been little but annual assessment work performed on any prospects in this district, and in 1912 the only pro-

[1] Louderback, G. D., op. cit., pp. 666–668.

[2] Browne, J. R., Mineral resources of the States and Territories west of the Rocky Mountains, 1867, p. 327.

[3] Idem, 1868, p. 316.

[4] Hague, Arnold, Mining industry: U. S. Geol. Expl. 40th Par., vol. 3, pp. 296–297, 1870.

[5] Burchard, H. C., Report of the Director of the Mint for 1881, p. 157.

ducer was the Red Metal Co. on the northwest side of Peavine Mountain. Some exploration work was being done on the Nevada Central ground by the Mason Valley Copper Co.

No estimates of the early production from the Peavine district are available, though it is said that several thousand dollars' worth of placer gold were recovered from what is now known as the Nevada Industrial placer. The ores from the veins near Poeville were treated in a stamp mill on the ground, but no data can be found on the production from this source.

The production from 1902 to 1908, inclusive, is given in the following table, compiled from the Mineral Resources volumes published by the United States Geological Survey. There was no recorded production from this district for the years 1909 to 1911, inclusive.

Production of the Peavine mining district, Washoe County, Nev.

	Ore treated.	Gold.	Silver.	Copper.	Total value.
	Tons.		*Ounces.*	*Pounds.*	
1902	50	$282	....	....	$282
1903	2,233	1,193	23,216	....	11,150
1904	....	....	....	....	....
1905	3	1,261	10	....	1,266
1906	412	5,045	1,000	....	5,783
1907	19	826	6	8,350	2,500
1908	141	2,791	86	4,197	3,390
	848	11,398	24,318	12,547	24,371

CHARACTER.

Ore deposition in the Peavine district seems to have taken place at two distinct periods. The earlier deposits are copper-gold veins and lenses in the schists and quartz monzonite. The later deposits are the replacement lodes in the andesites and possibly the similar deposit in quartz monzonite at the Nevada Central and Nevada Industrial mines.

COPPER-GOLD DEPOSITS.

In the vicinity of Poeville, in the schist near dikes of quartz monzonite porphyry, there are several quartz veins which carry pyrite and subordinate chalcopyrite that are said to be auriferous. These veins strike approximately parallel to the schistosity. At the Red Metal property lenses and small veins of bornite and oxidized copper minerals are found in siliceous schists lying nearly parallel to the structure which has a low dip to the southeast.

In both localities the inclosing rocks have not been intensely altered, though sericite is developed in some places near the walls of the Poeville veins, particularly in quartz monzonite porphyry.

SILVER-GOLD DEPOSITS IN ALTERED ANDESITES.

In the area mapped as Tertiary volcanics on Plate XIX there are many places where the augite andesite has been intensely altered by hot solutions. In some places this alteration has produced zones of hard siliceous rock with much quartz, which stand out as reefs above the surface. At other places the alteration has softened and bleached the rock over rather large areas. It is in the latter areas that the greater mineralization seems to have taken place. The alteration has resulted in masses of light-gray to white rock, composed of sericite, calcite, and quartz, with abundant disseminated pyrite and less widely scattered galena and sphalerite. In these zones there are stringers and nodules of quartz and the sulphides mentioned above, these minerals in some places being accompanied by silver sulphides and more rarely by silver chloride.

At the surface these areas of alteration are commonly stained a brilliant red by iron derived from the weathering pyrite, and at many places gypsum crystals are developed from the calcite and sulphuric acid formed from the breaking down of the pyrite.

In general these zones strike east, but near the east base of Peavine Mountain northward-trending zones of alteration are more common.

PYRITE GOLD ORES IN HYDROTHERMALLY ALTERED QUARTZ MONZONITE.

At the Nevada Central mine the ore occurs in an area of white bleached quartz monzonite. The rock is altered to sericite, calcite, and quartz, and contains abundant disseminated pyrite in fine grains. This material is cut by innumerable interlacing veinlets having a maximum width of 4 inches and consisting of 95 per cent pyrite and 5 per cent quartz.

The gold at the Nevada Industrial placer was apparently derived in large part from veinlets of pyrite that are found in softened bleached quartz monzonite.

The age of these deposits is not certain, but as the alteration is so much more intense than that of any of the deposits in the schists and associated quartz monzonite dikes, it is thought that the deposits were formed at the same time as the deposits in the andesite.

PROPERTIES.

Emma mine.—The Emma mine (No. 1, Pl. XIX, p. 184) is developed by an old abandoned shaft, said to be 300 feet deep, on the flat divide between the Lemmon Valley and Truckee Meadow drainage basins. The shaft is sunk on a vein in quartz monzonite which seems to strike east. The ore is said to have carried gold and silver.

Some fragments of ore found on the dump show a small amount of finely divided pyrite in a siliceous gangue.

Fravel mines.—The Fravel mines (No. 2, Pl. XIX) are located at Poeville and are not worked to any extent at the present time. At the tunnels northeast of town one man is keeping up the assessment work, but the old workings west of the cabins are now abandoned. These veins lie in schists cut by quartz monzonite porphyry dikes. These dikes in general conform to the schistosity which strikes N. 60° E., and at this place stand nearly vertical. The incline and tunnel west of town seem to be on a ledge that strikes north-northeast, about parallel to the schistosity. The work now being done northeast of the cabins is in a tunnel, running about S. 30° E., which, to judge from the dump, is quite extensive. On the surface there are several cuts on north-northeast zones of altered schists and quartz monzonite. The rock is quite soft and contains rather abundant, finely disseminated pyrite. This is cut by narrow stringers of quartz carrying abundant pyrite and a little chalcopyrite. The ore is said to have been generally low grade, carrying between $4.60 and $12 in gold and silver, and here and there a little copper. It is reported, however, that some small rich pockets of free milling ore were taken from the oxidized zones.

Some prospects near the summit of the ridge south of Poeville have been opened on similar ledges and are reported to show more galena and to have higher silver values than the Poeville ores.

Nevada Central mine.—The Nevada Central mine (No. 3, Pl. XIX) is 4¾ miles west-northwest of Reno at the base of Peavine Mountain. The ore body is a mass of hydrothermally altered, porphyritic quartz monzonite, impregnated with pyrite and a very little chalcopyrite, which is cut in all directions by stringers of pure pyrite and a little quartz. It lies near the contact of the quartz monzonite and the schists. The altered zone is at least 1,500 feet wide and one-half mile long, to judge from the outcrops, but all this zone does not appear to have suffered the same amount of mineralization. The workings consisted of a shaft of unknown depth and of a 600-foot crosscut tunnel. The first 400 feet of the tunnel, which runs N. 20° W., is in the partly consolidated sands and gravels of the Truckee formation and the last 200 feet is in the intensely sericitized and pyritized quartz monzonite porphyry.

Nevada Industrial placer.—The Nevada Industrial placer (No. 4, Pl. XIX) is on the northeast slope of Peavine Mountain, about 6 miles by road northwest of Reno. The placer ground occurs in the bottom of a small draw that cuts hydrothermally altered quartz monzonite porphyry intrusive into schists. The productive ground, to judge from the dumps, was about 1,500 feet long and 2 to 3 feet wide. Several tunnels and shafts are driven in the sides of the draw and show rather abundant disseminated pyrite in the soft bleached quartz monzonite. At the contacts of the schist and the intrusive there are narrow zones of magnetite, and small stringers of pyrite are seen in the altered igneous rock. East of the draw the alteration has produced silicified zones, usually bearing east and west, but the mineralization seems to have been less intense than in the softer sericitized areas. It is probable that this pyrite is slightly auriferous and that the concentration of this material by erosion can account, in large part, for the presence of the pay gravels. No workable veins have so far been discovered at the upper end of the productive ground, but small streaks of oxidized pyritic material are seen in the tunnels.

Recall prospect.—At the Recall prospect (No. 5, Pl. XIX) there are a 40-foot prospect shaft and several shallow pits on a north-south zone of hydrothermally altered augite andesite in which there is a small amount of disseminated pyrite.

Red Metal mine.—The Red Metal mine (No. 6, Pl. XIX) is 2 miles northwest of the summit of Peavine Mountain, 15 miles by road northwest of Reno and 7½ miles east-southeast of Purdy, Cal., its shipping point. The ores so far developed lie in rather small lenses parallel to the structure of feldspathic schists, probably derived from andesitic material. The main body consists of overlapping lenses, which occur through a vertical height of 50 feet and run from the surface for about 200 feet along the dip of 20° SE. The ore makes in crushed streaked schist, the crushing having resulted from movement nearly parallel to the schistosity. The property is developed by a tunnel which runs S. 34° E. for 598 feet, from which a raise of 50 feet was required to reach the ore. There are also some abandoned inclines from the surface on the ore zone. The ore is all more or less oxidized, though kernels of bornite, the chief sulphide seen, are being altered to chalcocite and copper pitch ore. Malachite and copper-iron sulphate were found in some ore, and in one specimen a thin film of light-blue copper phosphate was noted. This film is made up of radial groups of minute needles.

Reno Mispah.—The Reno Mispah (No. 7, Pl. XIX) is 2 miles north-northwest of Reno. At the surface there are a number of silicified andesite outcrops trending north and south, in some of which there is a little pyrite that is said to carry gold. There are several shallow pits and shafts in these zones, but the main development is a 600-foot crosscut tunnel headed under the largest cropping, which strikes N. 70° W. It attains a maximum depth of 150 feet. At this depth there is no silicification of the andesite, but at a few places narrow pyritized bleached andesite zones that have a northerly strike are cut. The back end of the tunnel is caved and, to judge from the dump, no silicified material was struck in the tunnel. The tunnel probably has not reached the point where the main zone of silicification should be cut.

Mazy mine.—At the Mazy mine of the Updike property (No. 8, Pl. XIX) there are two short tunnels and some surface work on a zone of crushed altered pyritized andesite which strikes N. 5°–10° W. and dips 50°–65° E. The tunnels are located in the bottom of the gulch and the shaft is about 30 feet above them on the east wall. Water stands in the shaft about 40 feet below the collar. The altered zone is 4 feet wide and at many places large crystals of gypsum are found embedded in the gougelike mass. The gypsum is probably derived from the weathering of the pyrite, which forms sulphuric acid, and this acid reacts with the calcite of the altered andesite to form the calcium sulphate. In the tunnels there are a few small veinlets of quartz with pyrite, galena, and sphalerite. Here and there small nodules of the same material are found in the gouge. This sulphide ore is said to carry $1.80 in gold and 6 ounces of silver to the ton.

Reno May.—The Reno May ground lies between the Updike and Reno Mispah. The vein, very similar to that of the Mazy, shows the same pyritized bleached andesite with small nodules of quartz sulphide ore. The only development consists of pits and shallow shafts and no bodies of ore of any size have been demonstrated.

Reno Rule.—The Reno Rule shaft is 3 miles northeast of Reno and 2 miles east of Fulton's quarry. It was not visited, though it was seen at a distance. It is sunk in the softened altered andesites to a reported depth of 400 feet. No ore of shipping grade is said to have been found.

Wedekind mine.—The Wedekind mine, north of Glendale (No. 9, Pl. XIX), was not visited in the course of this reconnaissance. The following notes are taken from a description of the property by H. C. Morris.[1] The so-called vein strikes northwest in hydrothermally altered andesite. The ore does not show at the surface in all places and most of the ledges are blind. The soft altered rock in the ore zones is heavily impregnated with gypsum and is very open. In general the ore looks like a ferruginous cemented conglomerate composed of imperfectly rounded pebbles. In the richer ore lead sulphides and carbonates carry free gold and silver sulphide and chloride. The ore zones are crossed by northerly striking "conglomerate" belts, which consist of altered andesite. The conglomeratic appearance is due to the partial zonal alteration of much-jointed

[1] Morris, H. C., Hydrothermal activity in the veins at Wedekind, Nev.: Eng. and Min. Jour., vol. 76, Aug. 22, 1903.

andesite, which has first been attacked at the corners and edges of the fragments by the mineralizing solutions. Dark quartz bands are found in the ore of the Desert King shaft near the streaks of rich sandy ore. The better ore occurs in small shoots and irregular chambers, most of them a foot in width but some of them swelling to 2 or 3 feet. In the Wedekind shaft, at a depth of 213 feet, a flow of warm water heavily charged with sulphureted hydrogen was cut, and the workings were hot. The Desert King and Bell shafts are quite dry. Silver and gold in some of the ore are said to be associated with lead and antimony.

INDEX.

G.

H.

I.

J.

K.

L.

M.

N.

O.

P.

Q.

R.

S.

NEVADA GHOST TOWNS & MINING CAMPS, by S. W. Paher. Large 8½ x 11 format, 492 pages, 700 illustrations. About 668 ghost towns are described with directions on how to get to them. It contains more pictures and describes more localities than any other Nevada book. Nearly every page brings new information and unpublished photos of the towns, the mines, the people and early Nevada life. It is the best seller of any book on Nevada ever published and it won the national "Award of Merit" for history. $30.00.

LAS VEGAS, AS IT BEGAN, AS IT GREW, by S. W. Paher. Large 9 x 12 format, 181 pages, 210 duotone illustrations, maps, index, biblio. In the 19th century Las Vegas was first a Spanish Trail waterhole, then a religious mission and then a ranch. Though founded in 1905 as a railroad town, it developed into a gambling center during distinct boom periods. The reader is led every step along these transitions in both the text and photos. $20.00

HISTORY OF NEVADA, 1540-1888, by H. H. Bancroft. 332 pp, 6 x 9. This is a reprint of the 1888 history of the early explorers through Nevada through the beginning of statehood in 1864 up to the date of publication. Large chapters discuss the Comstock Lode, state government, Indian wars, transportation, etc. $25.00.

HISTORY OF UTAH, 1840-1886, by H. H. Bancroft. 808 pp, illus., maps. Utah became a territory in 1850, three years after the Mormons came to settle Salt Lake Valley. Here is a detailed history of the early growth and development of Utah. Of interest to Nevadans is material on Carson County, western Nevada. $25.00.

HISTORY OF NEVADA, 1881, Thompson & West, ed. by Myron F. Angel. 900 pp, 7 x 10. This reprint of the 1881 history has 35 chapters covering every important development in Nevada's formative years — mining development, Indian raids, raucous politics, mail and press, early exploration, social organizations, etc. Granddaddy of Nevada historical works, and a researcher's must. $60.00.

INDEX TO HISTORY OF NEVADA, 1881, Thompson & West, by Helen Poulton. 160 pp, 7 x 10. This index of more than 10,000 entries is a most valuable auxiliary tool to the above history. This second edition adds a clear map. $8.00.

NEVADA, AN ANNOTATED BIBLIOGRAPHY, by S. W. Paher. 585 pp, 7 x 10, illus. Here is a researcher's guide and description of more than 2,544 books relating to the history and development of Nevada. The useful 74-page index has 3,550 subjects; 33 other bibliographies refer the researcher to countless Nevada materials. Printed on high quality paper and buckram bound. $95.00.

Write for catalog of more than 100 ghost town and treasure books on the Southwestern states.

PONDEROSA COUNTRY, by S. W. Paher. 48 pp, large 9 x 12 format, two-color illus, historic and modern. Color cover. Tours out of Reno guide the traveler to Virginia City, Carson City, Lake Tahoe, and ghost towns. Early Reno pictures included. Clothbound, $7.50.

CENTRAL ARIZONA GHOST TOWNS, by S. W. Paher. 48 pp, 9 x 12 map, 50 old-time photographs. Written in collaboration with Robert L. Spude, this book summarizes in text and pictures the mining camps between Phoenix and Prescott, including those in the Bradshaw Mountains and Wickenburg and Jerome areas. Maps and directions are a clear aid for ghosttowners. $3.95 paper; $7.50 cloth.

NORTHWESTERN ARIZONA GHOST TOWNS, by S. W. Paher. 32 pp, 9 x 12. Mining camps in the Las Vegas-Kingman area are described in a lively text with historic photographs. Oatman, El Dorado, Kingman, Searchlight, Chloride, and about 20 other ghost towns are included. $2.95.

COLORADO RIVER GHOST TOWNS, by S. W. Paher. 80 pp, full color cover, 50 old-time photographs, 20 pages of color, cloth (with dust jacket). $9.95.

CHLORIDE MINES AND MURALS, by Roy Purcell. 16 pp, 9 x 12. History of the silver camp of Chloride in northwestern Arizona is recalled by lifelong resident, with illustrations. Artist Purcell also interprets his Chloride murals. Color cover. $1.95.

TOMBSTONE, ARIZONA SILVER CAMP, by Bob Spude. 16 pp, 9 x 12. Author recounts the discovery, the boom days of the early 1880's (with illustrations), the mines, the town and its legendary violence. Color cover. $1.95.

DEATH VALLEY JUNCTION, THE STORY OF THE AMARGOSA OPERA HOUSE, by Marta Becket. Author tells her own story in 35 pictures (7 in color, including cover) of the development of an opera house complete with historic murals. $2.50.

DEATH VALLEY GHOST TOWNS, VOL. I, by S. W. Paher. 32 pp, 9 x 12, map, 50 old-time photos, color cover. Though Death Valley's mineral history is best known for its colorful eras of borax mining, there were two distinct periods of gold and silver mining also. The first one included Panamint and Calico, while the early 20th century boom produced Rhyolite, Greenwater, and others. About 35 mining camps are included. $2.95 paper.

DEATH VALLEY GHOST TOWNS, VOL. II, by S. W. Paher. 32 pp, 9 x 12. Mining camps of the Death Valley National Monument — Skidoo, Panamint City, and Old Stovepipe Wells — are joined by those immediately to the west, including Cerro Gordo, Darwin and Cartago. Essay on "The Prospector," the Tonopah & Tidewater Railroad, and desert driving hints. $2.95.

DEATH VALLEY GHOST TOWNS, by S. W. Paher. 48 pp, 9 x 12, map. The heart of the above two books combine to form this hardbound book. $7.50.

SOUTHWEST ARIZONA GHOST TOWNS, by S. W. Paher. 32 pp, 9 x 12. Within a day's drive of Yuma are important camps dating from the late 1850's: Gila City, LaPaz, Ehrenburg, and many more. There is information about the steamboats that plied the Colorado River upstream from Yuma. The colored cover etching of a placer mining camp; the back cover is an 1864 Bancroft map in color. $2.95.

SCOTTY'S CASTLE, DEATH VALLEY, by S. W. Paher. Large 9 x 12 format, 48 pp, heavily illustrated in color. Besides a history of Death Valley Scotty, there are many intricate pictures of the castle itself, with large captions to explain them. $3.75 paperback, with color cover.

THE CAMELS OF NEVADA, by Douglas McDonald. 32 pp, illus. In Nevada camel pack trains hauled salt, wood and even freight, also aiding early surveyors. But the beasts also brought problems. Modern camel races in Virginia City are recounted. $2.95 paperback.

THE BIG BONANZA, by Dan DeQuille (William Wright). 488 pages, illustrated, with intro. by Oscar Lewis. Indexed. Subtitled "An authentic account of the discovery, history, and working of the Comstock Lode," the Big Bonanza by DeQuille covers every phase of the epic rise of Virginia City, especially the special technology required to work the deep silver mines. $12.95.

GHOSTS OF THE GLORY TRAIL, by Nell Murbarger. 316 pages, illus. Indexed. Subtitled "Intimate glimpses into the past and present of 275 western ghosttowns," Ghosts of the Glory Trail is a fast-moving chronicle depicting the early-day mining stampedes. All Nevada counties are represented either in the 39 chapters on specific towns (such as Aurora, Rhyolite, Candelaria, Tuscarora, Delamar, etc.) or in the valuable ghost town directory containing 225 additional listings, some in eastern Calif. and western Utah. $12.95.

THE HISTORY OF NEVADA, by Sam P. Davis. 2 volumes. 1279 pages, illus. Reprint of the rare 1913 first edition. There are chapters on every phase of Nevada history including Indians, topography, the emigrants, mining, politics, journalism, education, religion, railroads, medicine, banking, agriculture, fraternal societies, drama, reclamation, divorce law, and a chapter on every Nevada county. $60.00.

NEVADA POST OFFICES, AN ILLUSTRATED HISTORY, by James Gamett and S. W. Paher. 160 pp, 7 x 10, illus. About 760 Nevada towns had post offices and here is a detailed list of them. There are chapters of the Pony Express, Wells-Fargo, and how to collect postal materials. 110 postal illustrations are in this research tool. Hardcover with color dust jacket. $30.00.

NEVADA GHOST TOWN TRAILS, by Mickey Broman. 80 pp, illus., maps. Quick reference and guide to 120 ghost towns. $2.95. Also available in same format is **CALIFORNIA GHOST TOWN TRAILS.**

GEM TRAILS IN CALIFORNIA, by Mickey Broman, 84 pp, illus., maps. Good maps lead traveler to numerous gem stone collecting areas. $2.95.

GOLD FEVER AND THE ART OF PANNING AND SLUICING, by Lois DeLorenzo. 80 pp, illus. A how-to book for the amateur. $3.50.

WHERE TO FIND GOLD AND GEMS IN NEVADA, by James Klein. 112 pp, maps, illus. Gold and gem sites in every county are included. $5.95.

THE WEEKEND GOLD MINER, by A. H. Ryan. 64 pp. A simple, easy-to-follow handbook. $1.95.

YOSEMITE TRAILS, by Lew and Ginny Clark. 144 pp, illus., maps, index. Complete guide to this National park. $5.95.

HIGH MOUNTAINS AND DEEP VALLEYS, by Lew and Ginny Clark. 191 pp, b&w and color illus., maps, index. Coverage is from Virginia City southward to all of eastern California (including Death Valley), and Tonopah-Goldfield-Beatty, Nevada. $6.95.

JOHN MUIR TRAIL COUNTRY, by Lew and Ginny Clark. 176 pp, illus., maps, index. A knowledgeable guide of the authors' beloved Sierra. $5.95.

COMSTOCK MINING AND MINERS, by Eliot Lord. 578 pp, illus., maps. Reprint of valuable 1883 mining report. $20.00.

RAILROADS OF NEVADA AND EASTERN CALIFORNIA, [Vol. 1, Northern Nevada; Vol. 2, Southern Nevada and Death Valley], by David F. Myrick. 469 and 492 pp, illus., maps, index. Valuable economic and railroad history with much politics and mining included. $30.00 each.

LOST LEGENDS OF THE SILVER STATE, by Gerald Higgs. Short stories on various Nevada topics. $8.95.